RESEARCH AND PERSPECTIVES IN ALZHEIMER'S DISEASE

Fondation Ipsen

Springer
Berlin
Heidelberg
New York
Barcelona
Budapest
Hong Kong
London
Milan
Paris
Santa Clara
Singapore
Tokyo

A.D. Roses K.H. Weisgraber
Y. Christen (Eds.)

Apolipoprotein E and Alzheimer's Disease

With 64 Figures and 10 Tables

Springer

Roses, A.D., M.D.
Joseph and Kathleen Bryan Alzheimer's Disease Research Center
Departments of Medicine (Neurology) and Neurobiology
P.O. Box 2900, Duke University Medical Center
Durham, NC 27710
USA

Weisgraber, K., M.D.
Gladstone Institute of Cardiovascular Disease
Department of Pathology
Cardiovascular Research Institute
University of California
San Francisco, CA 94141-9100
USA

Christen, Y., Ph.D.
Fondation IPSEN
24, rue Erlanger
75781 Paris
France

CIP data applied for

Die Deutsche Bibliothek - CIP-Einheitsaufnahme
Apolipoprotein E and Alzheimer's disease with tables/A D
Roses (ed) - Berlin, Heidelberg, New York, Barcelona,
Budapest, Hong Kong, London, Milan, Paris, Santa Clara,
Singapore, Tokyo Springer, 1996
(Research and perspectives in Alzheimer's disease)
ISBN-13: 978-3-642-80111-2 e-ISBN-13: 978-3-642-80109-9
DOI: 10.1007/ 978-3-642-80109-9
NE Roses, Allen D [Hrsg]

Softcover reprint of the hardcover 1st edition 1996

Cover design Springer-Verlag, Design & Production

Typesetting Best-set Typesetter Ltd , Hong Kong

SPIN 10518225 27/3136/SPS - 5 4 3 2 1 0 - Printed on acid-free paper

Preface

There is now considerable genetic evidence that the type 4 allele of the apolipoprotein E gene is a major susceptibility factor associated with late-onset Alzheimer's disease, the common form of the disease defined as starting after 60 years of age. The roles of apolipoprotein E in normal brain metabolism and in the pathogenesis of Alzheimer's disease are new and exciting avenues of research. This is why the Fondation Ipsen organised an international meeting on this topic in Paris on May 29, 1995, and the proceedings are contained in this book. The editors wish to thank Mrs Mary Lynn Gage for editorial and Mrs Jacqueline Mervaillie for the organization of the meeting in Paris.

Allen Roses
Karl Weisgraber
Yves Christen

Contents

Apolipoprotein E and Alzheimer's Disease: State of the Field After Two Years
A.D. Roses, *W.J. Strittmatter*, *A.M. Saunders*, *D.E. Schmechel*, and *M.A. Pericak-Vance* 1

Role of Apolipoprotein E in Alzheimer's Disease: Clues from its Structure
K.H. Weisgraber and *L.M. Dong* 11

Apolipoprotein E4, Cholinergic Integrity, Synaptic Plasticity and Alzheimer's Disease
J. Poirier, *M.-C. Delisle*, *R. Quirion*, *I. Aubert*, *J. Rocheford*, *I. Rousse*, *S. Gracon*, *M. Farlow*, and *S. Gauthier* 20

Pattern of Apolipoprotein E Immunoreactivity During Brain Aging
D.E. Schmechel, *M.O. Tiller*, *P. Tong*, *M. McSwain*, *S.-H. Han*, *R. Ange*, *D.S. Burkhart*, and *M.K. Izard* 29

Apolipoprotein E: Structure, Function, and Possible Roles in Modulating Neurite Extension and Cytoskeletal Activity
R.W. Mahley, *B.P. Nathan*, *S. Bellosta*, and *R.E. Pitas* 49

Apolipoprotein E Role in Maintaining the Integrity of the Aging Central Nervous System
E. Masliah, *M. Mallory*, *M. Alford*, *I. Veinbergs*, and *A.D. Roses* 59

Very Low Density Lipoprotein Receptor and Alzheimer's Disease
K. Ikeda, *T. Aizawa*, *S. Haga*, *N. Otsuka*, *M. Mizuguchi*, *Y. Namba*, *R. Machinami*, *S.U. Kim*, and *T. Yamamoto* 74

Apolipoprotein E and Apolipoprotein E Receptors in the Central Nervous System: Relationship with Alzheimer's Disease
B.T. Hyman, *G.W. Rebeck*, *R. Christie*, *K. Page*, *M. Irizarry*, and *D.K. Strickland* 97

Tau Protein and the Neurofibrillary Pathology of Alzheimer's Disease
M. Goedert 103

Tau Protein and Alzheimer Paired Helical Filament Assembly: Interacting Domains and Control Residues
E. Mandelkow, O. Schweers, J. Biernat, B. Trinczek, G. Drewes, and *E.-M. Mandelkow* 126

The βA4 Amyloid Peptide Enhances Cellular Uptake of Apolipoprotein E-Containing Lipoproteins by the Low-Density Lipoprotein Receptor-Related Protein/α_2-Macroglobulin Receptor
H. Scharnagl, K. Winkler, M. Hüttinger, D. Wittmann, M. Nauck, H. Wieland, W. Groß, T.G. Ohm, and *W. März* 136

Interactions of Apolipoprotein E with Microtubule-Associated Proteins
W.J. Strittmatter, V. de Serrano, D.Y. Huang, M.E. Martin, A.D. Roses, and *A.M. Saunders* 145

Chaperoning Amyloid in Alzheimer's Disease: The Art of Avoiding Sticky Situations?
B. Frangione, E.M. Castaño, F. Prelli, C. Soto, J. Ghiso, and *T. Wisniewski* 151

Apolipoprotein E Oxidation and Functional Consequences
B. Leininger-Muller, C. Jolivalt, T. Pillot, P. Lagrange, M.H. Livertoux, M.C. Grassiot, A. Minn, and *G. Siest* 161

Ethnic Variation in the Association Between APOE-ε4 and Alzheimer's Disease: A Follow-up Study
G. Maestre, P. Schofield, M. Chun, M.-X. Tang, B. Tycko, and *R. Mayeux* 170

The Search for Additional Alzheimer's Disease Genes
M.A. Pericak-Vance, P.M. Conneally, G.W. Small, A.M. Saunders, L. Yamaoka, P.C. Gaskell, C. Robinson, M. Ter-minassian, P.A. Locke, M. Pritchard, C.S. Haynes, J. Growdon, J.F. Gusella, A.D. Roses, and *J.L. Haines* 180

Subject Index 187

Contributors

Aizawa, T.
Department of Ultrastructure and Histochemistry, Tokyo Institute of Psychiatry, 2-1-8 Kamikitazawa, Setagaya, Tokyo, Japan

Alford, M.
University of California, San Diego, School of Medicine, Department of Neuroscience, La Jolla, CA 92093-0624, USA

Ange, R.
Departments of Medicine and Neurobiology, Joseph and Kathleen Bryan Alzheimer's Disease Research Center, P.O. Box 2900, Duke University Medical Center, Durham, NC 27710, USA

Aubert, I.
McGill Centre for Studies in Aging, Douglas Hospital Research Centre, 6875 La Salle Boulevard, Verdun, Quebec H4H 1R3, Canada

Bellosta, S.
Gladstone Institute of Cardiovascular Disease, Cardiovascular Research Institute, Department of Medicine, University of California, San Francisco, CA 94141-9100, USA

Biernat, J.
Max-Planck-Unit for Structural Molecular Biology, c/o DESY, Notkestr. 85, 22603 Hamburg, Germany

Burkhart, D.S.
Departments of Medicine and Neurobiology, Joseph and Kathleen Bryan Alzheimer's Disease Research Center, P.O. Box 2900, Duke University Medical Center, Durham, NC 27710, USA

Castano, E.M.
Department of Pathology, New York University Medical Center, 550 First Avenue, Room TH 427, New York, NY 10016, USA

Chang, P.
Departments of Medicine and Neurobiology, Joseph and Kathleen Bryan Alzheimer's Disease Research Center, P.O. Box 2900, Duke University Medical Center, Durham, NC 27710, USA

Christie, R.
Neurology Service, Massachusetts General Hospital, Boston, MA 02114, USA

Chun, M.
Gertrude H. Sergievsky Center and Department of Neurology, Columbia University, New York, NY 10032, USA

Conneally, P.M.
Department of Medical Genetics, Indiana University Medical Center, Indianapolis, IN 46223, USA

de Serrano, V.
Department of Medicine (Neurology), Joseph and Kathleen Bryan Alzheimer's Disease Research Center, P.O. Box 2900, Duke University Medical Center, Durham, NC 27710, USA

Delisle, M.-C.
McGill Centre for Studies in Aging, Douglas Hospital Research Centre, 6875 La Salle Boulevard, Verdun, Quebec H4H 1R3, Canada

Dong, L.-M.
Gladstone Institute of Cardiovascular Disease, Department of Pathology, Cardiovascular Research Institute, University of California, San Francisco, CA 94141-9100, USA

Drewes, G.
Max-Planck-Unit for Structural Molecular Biology, c/o DESY, Notkestr. 85, 22603 Hamburg, Germany

Farlow, M.
McGill Centre for Studies in Aging, Douglas Hospital Research Centre, 6875 La Salle Boulevard, Verdun, Quebec H4H 1R3, Canada

Frangione, B.
Department of Pathology, New York University Medical Center, 550 First Avenue, Room TH 427, New York, NY 10016, USA

Gaskell, P.C.
Departments of Medicine and Neurobiology, Joseph and Kathleen Bryan Alzheimer's Disease Research Center, P.O. Box 2900, Duke University Medical Center, Durham, NC 27710, USA

Gauthier, S.
McGill Centre for Studies in Aging, Douglas Hospital Research Centre, 6875 La Salle Boulevard, Verdun, Quebec H4H 1R3, Canada

Ghiso, J.
Department of Pathology, New York University Medical Center, 550 First Avenue, Room TH 427, New York, NY 10016, USA

Goedert, M.
MRC Laboratory of Molecular Biology, Hills Road, Cambridge CB2 2QH, UK

Gracon, S.
McGill Centre for Studies in Aging, Douglas Hospital Research Centre, 6875 La Salle Boulevard, Verdun, Quebec H4H 1R3, Canada

Grassiot, M.C.
Centre du Médicament, Université Henri Poincaré, CNRS URA 597, 30 Rue Lionnois, 54000 Nancy, and Centre de Médecine Préventive, 2 Rue du Doyen Jacques Parisot, 54500 Vandoeuvre, France

Groß, W.
Gustav Embden-Center of Biological Chemistry, Johann Wolfgang Goethe-University, 6000 Frankfurt, Germany

Growdown, J.
Department of Neurology, Massachusetts General Hospital, Boston, MA 02114, USA

Gusella, J.F.
Laboratory of Genetics and Aging, Massachusetts General Hospital, East Boston, MA 02129, USA

Haga, S.
Department of Ultrastructure and Histochemistry, Tokyo Institute of Psychiatry, 2-1-8 Kamikitazawa, Setagaya, Tokyo, Japan

Haines, J.L.
Laboratory of Genetics and Aging, Massachusetts General Hospital, East Boston, MA 02129, USA

Han, S.-H.
Departments of Medicine and Neurobiology, Joseph and Kathleen Bryan Alzheimer's Disease Research Center, P.O. Box 2900, Duke University Medical Center, Durham, NC 27710, USA

Haynes, C.S.
Departments of Medicine and Neurobiology, Joseph and Kathleen Bryan Alzheimer's Disease Research Center, P.O. Box 2900, Duke University Medical Center, Durham, NC 27710, USA

Huang, D.Y.
Department of Neurobiology, Joseph and Kathleen Bryan Alzheimer's Disease Research Center, P.O. Box 2900, Duke University Medical Center, Durham, NC 27710, USA

Hüttinger, M.
Department of Medical Chemistry, University of Vienna, Vienna, Austria

Hyman, B.T.
Neurology Service, Massachusetts General Hospital, Boston, MA 02114, USA

Ikeda, K.
Department of Ultrastructure and Histochemistry, Tokyo Institute of Psychiatry, 2-1-8 Kamikitazawa, Setagaya, Tokyo 156, Japan

Irizarry, M.
Neurology Service, Massachusetts General Hospital, Boston, MA 02114, USA

Izard, M.K.
Departments of Medicine and Neurobiology, Joseph and Kathleen Bryan Alzheimer's Disease Research Center, P.O. Box 2900, Duke University Medical Center, Durham, NC 27710, USA

Jolivalt, C.
Centre du Médicament, Université Henri Poincaré, CNRS URA 597, 30 Rue Lionnois, 54000 Nancy, and Centre de Médecine Préventive, 2 Rue du Doyen Jacques Parisot, 54500 Vandoeuvre, France

Kim, S.U.
Department of Neurology, University Hospital of UBC, 2211 Wesbrook Mall, Vancouver, British Columbia V6T 2B5, Canada

Lagrange, P.
Centre du Médicament, Université Henri Poincaré, CNRS URA 597, 30 Rue Lionnois, 54000 Nancy, and Centre de Médecine Préventive, 2 Rue du Doyen Jacques Parisot, 54500 Vandoeuvre, France

Leininger-Muller, B.
Centre du Médicament, Université Henri Poincaré, CNRS URA 597, 30 Rue Lionnois, 54000 Nancy, and Centre de Médecine Préventive, 2 Rue du Doyen Jacques Parisot, 54500 Vandoeuvre, France

Livertoux, M.H.
Centre du Médicament, Université Henri Poincaré, CNRS URA 597, 30 Rue Lionnois, 54000 Nancy, and Centre de Médecine Préventive, 2 Rue du Doyen Jacques Parisot, 54500 Vandoeuvre, France

Locke, P.A.
Laboratory of Genetics and Aging, Massachusetts General Hospital, East Boston, MA 02129, USA

Machinami, R.
Department of Pathology, University of Tokyo, Tokyo, Japan

Maestre, G.
Taub Center for Alzheimer's Disease Research in the City of New York, and Department of Pathology, Columbia University, New York, NY 10032, USA

Mahley, R.W.
Gladstone Institute of Cardiovascular Disease, Cardiovascular Research Institute, Departments of Pathology and Medicine, University of California, San Francisco, CA 94141-9100, USA

Mallory, M.
University of California, San Diego, School of Medicine, Department of Neuroscience, La Jolla, CA 92093-0624, USA

Mandelkow, E.
Max-Planck-Unit for Structural Molecular Biology, c/o DESY, Notkestr. 85, 22603 Hamburg, Germany

Mandelkow, E.-M.
Max-Planck-Unit for Structural Molecular Biology, c/o DESY, Notkestr. 85, 22603 Hamburg, Germany

Martin, M.E.
Department of Medicine (Neurology), Joseph and Kathleen Bryan Alzheimer's Disease Research Center, P.O. Box 2900, Duke University Medical Center, Durham, NC 27710, USA

Marz, W.
Division of Clinical Chemistry, Department of Medicine, Albert-Ludwigs University, Hugstetter Str. 55, 79106 Freiburg, Germany

Masliah, E.
University of California, San Diego, School of Medicine, Department of Neuroscience, La Jolla, CA 92093-0624, USA

Mayeux, R.
Gertrude H. Sergievsky Center and the Departments of Neurology and Pathology, Columbia University, New York, NY 10032, USA

McSwain, M.
Departments of Medicine and Neurobiology, Joseph and Kathleen Bryan Alzheimer's Disease Research Center, P.O. Box 2900, Duke University Medical Center, Durham, NC 27710, USA

Minn, A.
Centre du Médicament, Université Henri Poincaré, CNRS URA 597, 30 Rue Lionnois, 54000 Nancy, and Centre de Médecine Préventive, 2 Rue du Doyen Jacques Parisot, 54500 Vandoeuvre, France

Mizuguchi, M.
Department of Mental Retardation and Birth Defect, National Institute of Neuroscience, Tokyo, Japan

Namba, Y.
Department of Ultrastructure and Histochemistry, Tokyo Institute of Psychiatry, 2-1-8 Kamikitazawa, Setagaya, Tokyo, Japan

Nathan, B.P.
Gladstone Institute of Cardiovascular Disease, Cardiovascular Research Institute, Department of Medicine, University of California, San Francisco, CA 94141-9100, USA

Nauck, M.
Division of Clinical Chemistry, Department of Medicine, Albert-Ludwigs-University, Hugstetter Str. 55, 79106 Freiburg, Germany

Ohm, T.G.
Center of Morphology, Johann Wolfgang Goethe-University, 6000 Frankfurt, Germany

Otsuka, N.
Department of Pathology, University of Tokyo, Tokyo, Japan

Page, K.
Neurology Service, Massachusetts General Hospital, Boston, MA 02114, USA

Pericak-Vance, M.A.
Departments of Medicine and Genetics, Joseph and Kathleen Bryan Alzheimer's Disease Research Center, P.O. Box 2900, Duke University Medical Center, Durham, NC 27710, USA

Pillot, T.
Centre du Médicament, Université Henri Poincaré, CNRS URA 597, 30 Rue Lionnois, 54000 Nancy, and Centre de Médecine Préventive, 2 Rue du Doyen Jacques Parisot, 54500 Vandoeuvre, France

Pitas, R.E.
Gladstone Institute of Cardiovascular Disease, Cardiovascular Research Institute, Department of Pathology, University of California, San Francisco, CA 94141-9100, USA

Poirier, J.
McGill Centre for Studies in Aging, Douglas Hospital Research Centre, 6875 La Salle Boulevard, Verdun, Quebec H4H 1R3, Canada

Prelli, F.
Department of Pathology, New York University Medical Center, 550 First Avenue, Room TH 427, New York, NY 10016, USA

Pritchard, M.
Departments of Medicine and Neurobiology, Joseph and Kathleen Bryan Alzheimer's Disease Research Center, P.O. Box 2900, Duke University Medical Center, Durham, NC 27710, USA

Quirion, R.
McGill Centre for Studies in Aging, Douglas Hospital Research Centre, 6875 La Salle Boulevard, Verdun, Quebec H4H 1R3, Canada

Rebeck, G.W.
Neurology Service, Massachusetts General Hospital, Boston, MA 02114, USA

Robinson, C.
Departments of Medicine and Neurobiology, Joseph and Kathleen Bryan Alzheimer's Disease Research Center, P.O. Box 2900, Duke University Medical Center, Durham, NC 27710, USA

Rocheford, J.
McGill Centre for Studies in Aging, Douglas Hospital Research Centre, 6875 La Salle Boulevard, Verdun, Quebec H4H 1R3, Canada

Roses, A.D.
Departments of Medicine and Neurobiology, Joseph and Kathleen Bryan Alzheimer's Disease Research Center, P.O. Box 2900, Duke University Medical Center, Durham, NC 27710, USA

Rousse, I.
McGill Centre for Studies in Aging, Douglas Hospital Research Centre, 6875 La Salle Boulevard, Verdun, Quebec H4H 1R3, Canada

Saunders, A.M.
Department of Medicine, Joseph and Kathleen Bryan Alzheimer's Disease Research Center, P.O. Box 2900, Duke University Medical Center, Durham, NC 27710, USA

Scharnagl, H.
Division of Clinical Chemistry, Department of Medicine, Albert-Ludwigs-University, Hugstetter Str. 55, 79106 Freiburg, Germany

Schmechel, D.E.
Departments of Medicine and Neurobiology, Joseph and Kathleen Bryan

Alzheimer's Disease Research Center, P.O. Box 2900, Duke University Medical Center, Durham, NC 27710, USA

Schofield, P.
Gertrude H. Sergievsky Center and Department of Neurology, Columbia University, New York, NY 10032, USA

Schweers, O.
Max-Planck-Unit for Structural Molecular Biology, c/o DESY, Notkestrasse 85, 22603 Hamburg, Germany

Siest, G.
Centre du Médicament, Université Henri Poincaré, CNRS URA 597, 30 Rue Lionnois, 54000 Nancy, and Centre de Médicine Préventive, 2 Rue du Doyen Jacques Parisot, 54500 Vandoeuvre, France

Small, G.W.
Department of Psychiatry, University of California, 760 Werkwood Plaza, Los Angeles, CA 90024-1759, USA

Soto, C.
Department of Pathology, New York University Medical Center, 550 First Avenue, Room TH 427, New York, NY 10016, USA

Strickland, D.K.
American Red Cross, Bethesda, MD 20855, USA

Strittmatter, W.J.
Departments of Medicine (Neurology) and Neurobiology, Joseph and Kathleen Bryan Alzheimer's Disease Research Center, P.O. Box 2900, Duke University Medical Center, Durham, NC 27710, USA

Tang, M.-X.
Gertrude H. Sergievsky Center and Department of Neurology, Columbia University, New York, NY 10032, USA

Tanzi, R.E.
Laboratory of Genetic and Aging, Massachusetts General Hospital, East Boston, MA 02129, USA

Ter-minassian, M.
Laboratory of Genetic and Aging, Massachusetts General Hospital, East Boston, MA 02129, USA

Tiller, M.O.
Departments of Medicine and Neurobiology, Joseph and Kathleen Bryan Alzheimer's Disease Research Center, P.O. Box 2900, Duke University Medical Center, Durham, NC 27710, USA

Tong, P.
Departments of Medicine and Neurobiology, Joseph and Kathleen Bryan Alzheimer's Disease Research Center, P.O. Box 2900, Duke University Medical Center, Durham, NC 27710, USA

Trinczek, B.
Max-Planck-Unit for Structural Molecular Biology, c/o DESY, Notkestrasse 85, 22603 Hamburg, Germany

Tycko, B.
Center for Alzheimer's Disease Research in the City of New York, and Department of Pathology, Columbia University, New York, NY 10032, USA

Veinbergs, I.
University of California, San Diego, School of Medicine, Department of Neuroscience, La Jolla, CA 92093-0624, USA

Weisgraber, K.H.
Gladstone Institute of Cardiovascular Disease, Department of Pathology, Cardiovascular Research Institute, University of California, San Francisco, CA 94141-9100, USA

Wieland, H.
Division of Clinical Chemistry, Department of Medicine, Albert-Ludwigs-University, Hugstetter Str. 55, 79106 Freiburg, Germany

Winkler, K.
Division of Clinical Chemistry, Department of Medicine, Albert-Ludwigs-University, Hugstetter Str. 55, 79106 Freiburg, Germany

Wisniewski, T.
Department of Neurology, New York University Medical Center,
550 First Avenue, Room TH 427, New York, NY 10016, USA

Wittmann, D.
Gustav Embden-Center of Biological Chemistry, Johann Wolfgang
Goethe University, 6000 Frankfurt, Germany

Yamamoto, T.
Gene Research Center, Tohoku University, Sendai, Japan

Yamaoka, L.
Departments of Medicine and Neurobiology, Joseph and Kathleen Bryan
Alzheimer's Disease Research Center, P.O. Box 2900, Duke University
Medical Center, Durham, NC 27710, USA

Apolipoprotein E and Alzheimer's Disease: State of the Field After Two Years

A.D. Roses, W.J. Strittmatter, A.M. Saunders, D.E. Schmechel,* and *M.A. Pericak-Vance*

A Brief History of Apolipoprotein E and Alzheimer's Disease

In October 1992, at the Alzheimer's Disease Research Center Directors' meeting at the Society for Neurosciences (Anaheim, California), we reported that the apolipoprotein E-ε4 allele frequency was increased in late-onset familial Alzheimer's disease. There was no press release or news conference and, in fact, only one mention was made of these data by a reporter from the *Journal NIH Research* who attended the meeting. The first published paper appeared five months later in March 1993, 26 months before this IPSEN Fondation conference (Strittmatter et al. 1993a). In April 1993, during a 10-minute platform presentation at the American Academy of Neurology meeting, the association of APOE4 with 176 autopsy-confirmed cases of sporadic late-onset Alzheimer disease was presented (Saunders et al. 1993). There was no press release or public announcement for these data, which remained in press until August 1993. Because the Academy meeting was held in New York City, it was attended by a local medical science reporter from the Wall Street Journal who was impressed with the spirited criticism during the five minute question period and the hallway discussion that followed.

On June 7, 1993, a front page article in the *Wall Street Journal* appeared and provided the impetus that jump-started confirmations of this association throughout the world. Two months later, when the Saunders et al. (autopsy-confirmed sporadic series) paper appeared in *Neurology*, simultaneously with the later submitted Corder et al. family-based epidemiological analysis, many more laboratories became informed about the association (Corder et al. 1993; Saunders et al. 1993). Over the next 21 months, until this meeting, more than 90 articles, letters, or abstracts have been published by research groups throughout the world, confirming the increased allele frequency of APOE4 with Alzheimer's disease (Roses 1994; Roses 1995b; Roses et al. 1994). To many investigators, the role of the polymorphisms at the APOE

*Departments of Medicine (Neurology) and Neurobiology, Joseph and Kathleen Bryan Alzheimer's Disease Research Center, P.O. Box 2900, Duke University Medical Center, Durham, NC 27710, USA

A D Roses et al (Eds)
Apolipoprotein E and Alzheimer's Disease

locus as genetic susceptibility factors in Alzheimer's disease appeared surprising and, literally, unbelievable. In a field where the relevant causative hypotheses were still derived from constituents of the plaques and tangles that Alois Alzheimer described in 1907, the rapid and universal confirmations changed the focus of research from neuropathology of amyloid plaques and neurofibrillary tangles to the isoform-specific metabolism of apoE.

In fact, the discovery of the role of APOE in Alzheimer's disease developed directly out of a long-term positional cloning strategy for genetic factors in familial Alzheimer's disease that began in 1981 (Pericak-Vance et al. 1988). The discovery of the dose-related association of APOE4 with Alzheimer's disease was the first description of a susceptibility locus for a common late-onset disease found by using these positional cloning or "reverse genetic" methods (Roses 1995c). The linkage strategy was based on defining the chromosomal location of a gene for late-onset familiar Alzheimer's disease, without concern for any prevalent phenotypic hypothesis. Candidate gene considerations flow from the identification of a smaller linkage region than the whole human genome.

In 1987, when the amyloid precursor protein gene was localized to chromosome 21 and four early-onset, autosomal dominant Alzheimer's disease families were apparently linked to chromosome 21, the coincidence was greeted warmly as dogma (Goldgaber et al. 1987; St George-Hyslop et al. 1987; Tanzi et al. 1987) Data that could not confirm the chromosome 21 linkage were virtually ignored (Pericak-Vance et al. 1988; Schellenberg et al. 1988). When the linkage of late-onset Alzheimer's disease to chromosome 19 was published in 1991, it "was dismissed by many observers," with virtually no other laboratory searching for an Alzheimer's disease gene on chromosome 19 (Lander and Schork 1994; Pericak-Vance et al. 1991). When missense mutations in the APP gene were found later in 1992, it was immediately assumed that they would be present in most of familial Alzheimer disease (Goate et al. 1991; Pericak-Vance et al. 1991). To date, less than 20 AD or mixed hemorrhage/AD families with APP mutations have been identified worldwide. With the publication of the chromosome 14 linkage for most of the other large, early-onset Alzheimer disease families, including the four originally reported on chromosome 21, the heterogeneity of Alzheimer's disease became well-established (Sherrington et al. 1995; St George-Hyslop et al. 1987). It is important to realize that the Ipsen Fondation series of conferences on Alzheimer's disease has documented these trends well, starting with the second Alzheimer's disease meeting in March 1988. This detailed published series makes interesting reading in 1995, providing historical insight into the obstinate influence of the amyloid hypotheses on genomic searches over the past decade (Selkoe 1994). This unusually strong bias was not a factor in the gene mapping of most other diseases, where the gene mapping data dictated candidate genes rather than a single candidate, APP, overwhelming the search.

Summary of the Apolipoprotein E Data

The genetic association of APOE4 as a susceptibility gene or risk factor increasing the risk and lowering the age of onset distribution of Alzheimer disease is now well established (Corder et al. 1994; Corder et al. 1993). The APOE4 allele frequency is highly significantly increased in affected individuals compared to age- and sex-matched controls in familial and sporadic late-onset Alzheimer's disease. More than 90 publications, letters, and abstracts have confirmed this association in less than two years (Roses 1995b). The inheritance of each APOE4 allele confers increased risk of disease expression and a younger age of onset in a dose-dependent manner (Corder et al. 1993). The effect of the more uncommon allele, APOE2, is to decrease the risk and increase the age of onset compared to the common allele APOE3 (Corder et al. 1994; Roses 1994, 1995a). Thus the effect of inheriting APOE4 genotypes leads to a younger distribution of Alzheimer's disease expression. The influence of apoE isoforms on the age of onset also is present in several large, early-onset APP-mutation Alzheimer's disease families (Alzheimer's Disease Collaborative Group 1993; St George-Hyslop et al. 1994). No apparent interaction occurs in the very early-onset families carrying missense mutations of the S182 gene on chromosome 14 (Sherrington et al. 1995; Van Broeckhoven et al. 1994). There is no evidence that the mechanisms of pathogenesis involving the S182 missense mutations and that of late-onset Alzheimer's disease are similar; rather they are only analogies based on similar clinical and neuropathological phenotypes (Roses 1994; Roses et al. 1995; Selkoe 1994; Sherrington et al. 1995).

The analysis of APOE genotypes and duration of disease from onset to autopsy also predicts neuropathological variations observed in patients. (Rebeck et al. 1993; Schmechel et al. 1993). Antibodies to apoE are immunoreactive in amyloid plaques, neurofibrillary tangles, and vascular amyloid depositions in Alzheimer's disease patients (Namba et al. 1991; Strittmatter et al. 1993a). When brain sections from patients who are homozygous for APOE4 are compared to those of APOE3-homozygous patients of similar duration of illness, the former have larger and more dense Aβ immunoreactivity (Schmechel et al. 1993). In vitro, apoE4 also binds Aβ peptide more rapidly than apoE3, thus providing a test tube correlation of the phenotypic neuropathological observations (Strittmatter et al. 1993b).

ApoE antibodies also stain neurons in autopsy brains from elderly controls, Alzheimer's disease' patients, and aged primates (Han et al. 1994a; Han et al. 1994b; Strittmatter et al. 1993a). In Alzheimer's disease patients, many apoE-immunoreactive neurons also contain immunoreactive neurofibrillary tangles. However, in situ hybridization studies suggest that apoE mRNA is not usually expressed in neurons (Diedrich et al. 1991). Immunoelectron microscopic studies have clearly localized apoE in the cytoplasm of neurons (Han et al. 1994a). To account for an apparent intraneuronal metabolic role for apoE, we have considered the hypothesis that one of the normal functions of apoE within neurons is to contribute to the stabilization of microtubular integrity (Strittmatter et al. 1994b). We have demonstrated that apoE3 binds non-phosphorylated tau in vitro to form an SDS-resistant complex, but here is virtually no binding of apoE4 to tau under similar conditions. Neither apoE3 nor apoE4 bind to hyperphosphorylated tau (Strittmatter et al. 1994a). ApoE3 binding occurs at the microtubule-binding repeat domains of the tau molecule. ApoE3 can bind to synthetic peptides of each of the four microtubule binding domains of tau, and can be abolished by the phosphorylation of ser^{262} in domain 1 (Huang et al. 1995). Similar apoE isoform-specific interactions with the other major microtubular associated protein, MAP2c, have also been demonstrated (Huang et al. 1994).

We have proposed a model to explain the increased rate of disease expression as a function of inheritance of APOE4 allele(s). We have suggested that apoE3 (and/or apoE2) plays a protective role by binding and sequestering free tau, perhaps by facilitating microtubule stabilization and decreasing the rate of paired helical filament formation with subsequent phosphorylation (Roses et al. 1995; Strittmatter et al. 1994b). We are also studying the apoE allele-specific interactions with tau, β-tubulin and microtubules, as well as the kinetics of tau phosphorylation in vitro, and the intracellular trafficking of apoE in tissue cell systems. We are also modeling the molecular interactions of apoE3/apoE2 with tau and MAP2 to consider a molecular mimic as a potential therapeutic agent.

Rationale for this Conference

In the two years since the original publication of the apoE data, there has been considerable progress and activity in many laboratories throughout the world. This IPSEN Fondation meeting is the first public meeting focusing on the role of apoE in the pathogenesis of Alzheimer's disease. There are approximately as many hypotheses as there are scientists thinking about the problem. The participants who were invited as speakers were scientists who had already published on the structure and function of apoE, the role of apoE in the central nervous system, or the genetics of APOE and Alzheimer's disease through 1993. Their charge was to bring the field up-to-date on their research. The abstracts provided an opportunity to see work as yet unpublished from many other laboratories. These include studies confirming apoE localization in neuronal cytoplasm, mechanisms of apoE/Aβ interactions and intracellular metabolism, behavioral studies of apoE-deficient mice, isoform-specific involvement of apoE in Aβ fibrillogenesis, and the relationship of APOE genotype to drug therapeutic responses.

We also had the opportunity to critically review and discuss some of the data reaching the literature, to weight the data with respect to the conclusions reached in those papers, and to discuss those areas that seem somewhat unclear in 1995. In some cases, far-reaching conclusions have been based on minimal data or on data with highly selected ascertainment biases (Lannfelt et al. 1994; Mayeux et al. 1993). Areas of discussion in the chapters include, but are not limited to: (a) the effect of APOE4 and APOE2 in the APP and CH14 (S182) forms of early-onset Alzheimer's disease; (b) the allele frequencies in African-Americans and other racial and ethnic groups; (c)the specificity of the APOE4 association for Alzheimer's disease and other dementing diseases; (d) the role of apoE "testing" for differential diagnosis or predictions (Roses 1995b); (e) theories of pathobiology involving amyloid deposition, neurofibrillary tangle formation, paired helical filament formation, etc.; (f) animal models such as the apoE-deficient mice, human apoE isoforms transgenic in apoE-deficient backgrounds (Masliah et al., submitted for publication), APP717-derived transgenic mice (Games et al. 1995), human tau transgenics, etc.; and (g) other potential susceptibility genes for late-onset disease.

In fact, some of the prevalent arguments of early 1994 – concerning: (a) linkage disequilibrium with a different, unknown gene near APOE on chromosome 19 (Chartier-Harlin et al. 1994; Liddell et al. 1994; Tsuda et al. 1994; Yu et al. 1994); (b) the validity of in vitro and in vivo apoE interactions with Aβ or tau; (c) the localization of apoE in human neurons; and (d) the disposition of apoE receptors in the central nervous system – were addressed in the literature in 1994 and will be expanded in 1995. This IPSEN Conference attempts to describe the directions that research on apoE in Alzheimer disease is moving in the laboratories of the speakers and the poster contributors in 1995, including animal models, clinical applications, epidemiology, cell biology, biochemistry, and pharmacology. The genetic relationship of APOE genotype to the rate (age of onset distribution) of Alzheimer's disease expression has been confirmed, setting the stage for hypothesis-driven experimentation to test the biological relationship of apoE isoforms to theories of pathogenesis.

State of the Field

Confirmations of the allele frequency of APOE4 and Alzheimer's disease appeared rapidly, requiring only DNA from patients and controls for PCR or serum for isoelectric focusing. Immunocytological confirmations of intraneuronal apoE in controls and Alzheimer's disease patients, disposition of LDL, LRP, and VLDL receptors on the surface of various cells in the central nervous system, and immunoelectron microscopy of apoE in human, primate, and other animal neurons are in progress or in press from a number of neurobiology laboratories. Preliminary data from several laboratories have examined whether low level mRNA expression of the APOE gene occurs in neurons and neuronal culture systems. There is also considerable research activity characterizing the role of apoE receptors in endocytosis and other potential mechanisms of apoE entry into cells.

One such study, reported as a poster by Schnargl et al. at this conference, appears as a contribution to this volume. The research may be of major importance because it not only speaks to the issue of intracellular apoE trafficking, but provides a putative normal metabolic function for the Aβ protein.

Ikeda et al. (this volume) have characterized the cellular disposition of VLDL receptors in the brain. The VLDL receptor was cloned in 1992, the year after Professor Ikeda and his colleagues reported apoE immunoreactivity of plaques and tangles in Alzheimer's disease brain (Namba et al. 1991; Sakai et al. 1994). The VLDL receptor is present on the surface of glia and neurons, providing further diversity to the potential isoform-specific metabolism of apoE in brain. In primate studies, Schmechel et al. (this volume) provided evidence that apoE can be observed intracellularly in neurons throughout adult life, with numbers of positively stained cells increasing with age.

Specific isoforms of apoE form an SDS-insoluble complex with several proteins associated with the pathology of Alzheimer's disease. Strittmatter and colleagues have demonstrated that apoE4 forms a complex with Aβ more rapidly than does apoE3, and that apoE3 forms an isoform-specific SDS-insoluble complex with tau and MAP2c (Huang et al. 1994, 1995; Strittmatter et al. 1993a,b, 1994a). These irreversible interactions have defied equilibrium studies, but represent an interesting phenomenon of apoE isoform-specific binding. Surface plasmon resonance techniques have been used to examine real-time molecular association and dissociation of apoE2, apoE3 and apoE4 with tau. These studies offer the possibility of measuring kinetics, affinities, and molecular structural interactions that may provide insight into a unique role for apoE isoform-specific mechanisms of protein interactions. Strittmatter and colleagues introduced the method and preliminary data during this conference.

The elegant detailed structural analysis of the molecular differences between the apoE isoforms stressed the salt bridge bonding of Arg61 as the critical variable in apoE interactions (Dong et al. 1994). Most investigators have focused on the Cys/Arg variability at positions 112 and 158, assuming the involvement of putative disulfide bonds. Weisgraber and colleagues have demonstrated that the uniquely human residue, Arg61, may be the critical element for isoform-specific interactions in humans.

Mahley and colleagues expanded their studies of the effect of apoE on neuronal cell culture systems, emphasizing isoform-specific effects on neuronal outgrowth and branching (Nathan et al. 1994). New data supported a differential effect of apoE3 and apoE4 transfections on β-tubulin and the cytoskeletal system of neuronal cells in culture.

Masliah provided evidence for age-dependent synaptic simplification in apoE deficient mice, with decreased MAP2 and synaptophysin immunoreactivity over the first year (Masliah et al. 1995). Intracerebral injections of apoE partially restored synaptic simplification. Electron microscopy of dendrites illustrated ballooning with fragmentation of the endomembrane systems (Roses 1995a).

Poirier looked at the apparent responsiveness of Alzheimer disease patients to Tacrine as a function of their APOE genotype. Studies of a series of responders and non-responders provided by Parke-Davis further suggested that APOE4 carriers responded significantly less to the drug. Several poster presentations also examined the potential use of APOE genotyping as a measure of drug responsiveness.

Pericak-Vance and colleagues presented the current status of genomic screening for additional late-onset Alzheimer's disease susceptibility loci. Data reaching empirically established significance levels for two additional genomic loci, based on Affected Pedigree Member analyses, sib-pair analyses and LOD score methods, were presented. Delineation of the additional chromosomal loci awaits confirmation experiments using a separate group of Alzheimer's disease families prospectively separated from those used for screening for purposes of independent confirmation.

Initial discussion concerning the possibility of linkage disequilibrium accounting for the association of APOE4 and Alzheimer's disease has found no support in any data set. A remarkable aspect of the genetic association has been its rapid and widespread confirmation, possible because the PCR methods and the collections of DNA from patients and controls are readily available.

There are multiple reports that apoE4 accelerates Aβ fibril formation and leads to more and denser amyloid immunostained plaques in APOE4 carriers, including those with Alzheimer's disease (Ma et al. 1994; Sanan et al. 1994; Schmechel et al. 1993; Wisniewski et al. 1994). The relationship of Aβ deposition to the causation of neuronal cell death in Alzheimer disease remains hypothetical (Selkoe 1994).

Perhaps the most important neurobiological finding to result from the genetic data is the immunolocalization of apoE protein in the cytoplasm, endosomes, and peroxisomes of neurons in Alzheimer's disease and controls (Han et al. 1994a). Since APOE mRNA is not known to be expressed in neurons, the presence of very low concentrations of intracellular apoE suggests a tightly regulated transport mechanism that can deliver very small amounts of apoE into the cytoplasm of the cell. There may be low level APOE expression but, to date, no data for this explanation have been demonstrated.

Isoform-specific interactions with tau and MAP2c suggest a potential role in microtubule stability or repair mechanisms. These hypotheses are now being tested. There are promising therapeutic possibilities offered by improving microtubular stability with a mimic of the putative beneficial effect of apoE2, or inhibiting a putative negative influence of apoE4.

Finally the application of APOE testing to clinical differential diagnosis of dementia and clinical research as a biologic risk factor will be further developed as more epidemiologic data reach publication. Current data suggest that APOE alleles may play a role in the prediction of age-dependent memory problems and responses to some medications, but not predictions of age of onset of Alzheimer's disease for an individual (Roses 1995b).

References

Alzheimer's Disease Collaborative Group (1993) Apolipoprotein E genotype and Alzheimer's disease Lancet 342 737–738

Chartier-Harlin MC, Parfitt M, Legrain S, Pérez-Tur J, Brousseau T, Evans A, Berr C, Vidal O, Roques P, Gourlet V, Fruchart JC, Delacourte A, Rossor M, Amouyel P (1994) Apolipoprotein E, ε4 allele as a major risk factor for sporadic early- and late-onset forms of Alzheimer's disease: Analysis of the 19q13 2 chromosomal region. Hum Mol Genet 3. 567–574

Corder EH, Saunders AM, Strittmatter WJ, Schmechel DE, Gaskell PC, Small GW, Roses AD, Haines JL, Pericak-Vance MA (1993) Gene dose of apolipoprotein E type 4 allele and the risk of Alzheimer's disease in late onset families Science 261 921–923

Corder EH, Saunders AM, Risch NJ, Strittmatter WJ, Schmechel DE, Gaskell PC Jr, Rimmler JB, Locke PA, Conneally PM, Schmader KE, Small GW, Roses AD, Haines JL, Pericak-Vance MA (1994) Protective effect of apolipoprotein E type 2 allele for late onset Alzheimer disease Nat Genet 7. 180–184

Diedrich JF, Minnigan H, Carp RI, Whitaker JN, Race R, Frey W, Hazse AT (1991) Neuropathological changes in scrapie and Alzheimer's disease are associated with increased expression of apolipoprotein E and cathepsin D in astrocytes. J Virol 65 4759–4768

Dong L-M, Wilson C, Wardell MR, Simmons T, Mahley RW, Weisgraber KH, Agard DA (1994) Human apolipoprotein E. Role of arginine 61 in mediating the lipoprotein preferences of the E3 and E4 isoforms. J Biol Chem 269 22358–22365

Games D, Adams D, Alessandrini R, Barbour R, Berthelette P, Blackwell C, Carr T, Clemens J, Donaldson T, Gillespie F, Guido T, Hagopian S, Jonhson-Wood K, Khan K, Lee M, Leibowitz P, Lieberburg I, Little S, Masliah E, McConlogue L, Montoya-Zavala M, Mucke L, Paganini L, Penniman E, Power M, Schenk D, Seubert P, Snyder B, Soriano F, Tan H, Vitale J, Wadsworth S, Wolozin B, Zhao J (1995) Alzheimer-type neuropathology in transgenic mice overexpressing V717F β-amyloid precursor protein. Nature 373: 523–527

Goate A, Chartier-Harlin M-C, Mullan M, Brown J, Crawford F, Fidani L, Giuffra L, Haynes A, Irving N, James L, Mant R, Newton P, Rooke K, Roques P, Talbot C, Pericak-Vance M, Roses A, Williamson R, Rossor M, Owen M, Hardy J (1991) Segregation of a missense mutation in the amyloid precursor protein gene with familial Alzheimer's disease Nature 349: 704–706

Goldgaber D, Lerman MI, McBride OW, Saffiotti U, Gajdusek DC (1987) Characterization and chromosomal localization of a cDNA encoding brain amyloid of Alzheimer's disease Science 235. 877–880

Han S-H, Einstein G, Weisgraber KH, Strittmatter WJ, Saunders AM, Pericak-Vance M, Roses AD, Schmechel DE (1994) Apolipoprotein E is localized to the cytoplasm of human cortical neurons: a light and electron microscopic study J Neuropath Exp Neurol 53: 535–544

Han S-H, Hulette CM, Saunders AM, Einstein G, Pericak-Vance MA, Strittmatter WJ, Roses AD, Schmechel DE (1994) Apolipoprotein E is present in hippocampal neurons without neurofibrillary tangles in Alzheimer's disease and in age-matched controls. Exper Neurol 128. 13–26

Huang DY, Goedert M, Jakes R, Weisgraber KH, Garner CC, Saunders AM, Pericak-Vance MA, Schmechel DE, Roses AD, Stritmatter WJ (1994) Isoform specific interactions of apolipoprotein E with the microtubule-associated protein MAP2c. implications for Alzheimer's disease Neurosci Lett 182 55–58

Huang DY, Weisgraber KH, Goedert M, Saunders AM, Strittmatter WJ (1995) ApoE3 binding to tau tandem repeat I is abolished by tau serine_{262} phosphorylation Neurosci Lett 192: 1–4

Lander ES, Schork NJ (1994) Genetic dissection of complex traits Science 265. 2037–2048

Lannfelt L, Lilius L, Nastase M, Viitanen M, Fratiglioni L, Eggertsen G, Berglund L, Angelin B, Linder J, Winblad B, Basun H (1994) Lack of association between apolipoprotein E allele ε4 and sporadic Alzheimer's disease. Neurosci Lett 169: 175–178

Liddell M, Williams J, Bayer A, Kaiser F, Owen M (1994) Confirmation of association between the ε4 allele of apolipoprotein E and Alzheimer's disease J Med Genet 31: 197–200

Ma J, Yee A, Brewer HJ, Das S, Potter H (1994) Amyloid-associated proteins α1-antichymotrypsin and apolipoprotein E promote assembly of Alzheimer β-protein into filaments. Nature 372 92–94

Masliah E, Mallory M, Ge N, Alford M, Veinbergs I, Roses AD (1995) Neurodegeneration in the central nervous system of apoE-deficient mice Exp Neurol 136 107–122

Mayeux R, Stern Y, Ottman R, Tatemichi TK, Tang M-X, Maestre G, Ngai C, Tycko B, Ginsberg H (1993) The apolipoprotein ε4 allele in patients with Alzheimer's disease. Ann Neurol 34 752–754

Namba Y, Tomonaga M, Kawasaki H, Otomo E, Ikeda L (1991) Apolipoprotein E immunoreactivity in cerebral amyloid deposits and neurofibrillary tangles in Alzheimer's disease and kuru plaque amyloid in Creutzfeldt-Jakob disease Brain Res 541: 163–166

Nathan BP, Bellosta S, Sanan DA, Weisgraber KH, Mahley RW, Pitas RE (1994) Differential effects of apolipoproteins E3 and E4 on neuronal growth in vitro. Science 264: 850–852

Pericak-Vance MA, Bebout JL, Gaskell PC, Yamaoka LH, Hung W-Y, Alberts MJ, Walker AP, Bartlett RJ, Haynes CA, Welst KA, Earl NL, Heymark A, Clark CM, Roses AD (1991) Linkage studies in familial Alzheimer's disease: evidence for chromosome 19 linkage. Am J Hum Genet 48 1034–1050

Pericak-Vance MA, Yamaoka LH, Haynes CS, Speer MC, Haines JL, Gaskell PC, Hung W-Y, Clark CM, Heyman AL, Trofatter JA, Eisenmenger JP, Gilbert JR, Lee JE, Alberts MJ, Dawson DV, Bartlett RJ, Earl NL, Siddique T, Vance JM, Conneally PM, Roses AD (1988) Genetic linkage studies in Alzheimer's disease families. Exp Neurol 102 271–279

Rebeck GW, Reiter JS, Strickland DK, Hyman BT (1993) Apolipoprotein E in sporadic Alzheimer's disease allelic variation and receptor interactions. Neuron 11: 575–580

Roses AD (1994) Apolipoprotein E affects the rate of Alzheimer disease expression: beta-amyloid burden is a secondary consequence dependent on APOE genotype and duration of disease. J Neuropathol Exp Neurol 53 429–437

Roses AD (1995a) Alzheimer disease as a model of molecular gerontology J NIH Res 7: 51–57

Roses AD (1995b) Apolipoprotein E genotyping in differential diagnosis, not prediction, of Alzheimer's disease Ann Neurol 38:6–14

Roses AD (1995c) Perspective: On the metabolism of apolipoprotein E and the Alzheimer diseases. Exp Neurol 132 149–156

Roses AD, Strittmatter WJ, Pericak VM, Corder EH, Saunders AM, Schmechel DE (1994) Clinical application of apolipoprotein E genotyping to Alzheimer's disease. Lancet 343 1564–1565

Roses AD, Einstein E, Gilbert J, Goedert M, Han S-H, Huang D, Hulette C, Masliah E, Pericak-Vance MA, Saunders AM, Schmechel DE, Strittmatter WJ, Weisgraber KH, Xi P-T (1995) Morphological, biochemical, and genetic support for an apolipoprotein E effect on microtubular metabolism. Ann NY Acad Sci, in press

Sakai J, Hoshino A, Takahashi S, Miura Y, Ishii H, Suzuki H, Kawarabayasi Y, Yamamoto T (1994) Structure, chromosome location, and expression of the human very low density lipoprotein receptor gene. J Biol Chem 269: 2173–2182

Sanan DA, Weisgraber KH, Russell SJ, Mahley RW, Huang D, Saunders A, Schmechel D, Wisniewski T, Frangione B, Roses AD, Strittmatter WJ (1994) Apolipoprotein E associates with β amyloid peptide of Alzheimer's disease to form novel monofibrils. Isoform apoE4 associates more efficiently than apoE3. J Clin Invest 94: 860–869

Saunders AM, Strittmatter WJ, Schmechel D, St George-Hyslop PH, Pericak-Vance MA, Joo SH, Rosi BL, Gusella JF, Crapper-MacLachlan DR, Alberts MJ, Hulette C, Crain B, Goldgaber D, Roses AD (1993) Association of apolipoprotein E allele ε4 with late-onset familial and sporadic Alzheimer's disease. Neurology 43: 1467–1472

Schellenberg GD, Bird TD, Wijsman EM, Moore DK, Boehnke M, Bryant EM, Lampe TH, Nochlin D, Sumi SM, Deeb SS, Beyreuther K, Martin GM (1988) Absence of linkage of chromosome 21q21 markers to familiar Alzheimer's disease Science 241 1507–1510

Schmechel DE, Saunders AM, Strittmatter WJ, Crain BJ, Hulette CM, Joo SH, Pericak-Vance MA, Goldgaber D, Roses AD (1993) Increased amyloid β-peptide depositon in cerebral cortex as a consequence of apolipoprotein E genotype in late-onset Alzheimer disease Proc Natl Acad Sci USA 90 9649–9653

Selkoe DJ (1994) Alzheimer's disease a central role for amyloid J Neuropathol Exp Neurol 53 438–447

Sherrington R, Rogaev EI, Liang Y, Rogaeva EA, Levesque G, Ikeda M, Chi H, Lin C, Li G, Holman K, Tsuda T, Mar L, Foncin J-F, Bruni AC, Montesi MP, Sorbi S, Rainero I, Pinessi L, Nee L, Chumakov I, Pollen D, Brookes A, Sanseau P, Polinsky RJ, Wasco W, Da Silva HAR, Haines JL, Pericak-Vance MA, Tanzi RE, Roses AD, Fraser PE, Rommens JM, St George-Hyslop PH (1995) Cloning of a gene bearing missense mutations in early-onset familial Alzheimer's disease Nature 375 754–760

St George-Hyslop P, Tanzi R, Polinsky R, Haines JL, Nee L, Watkins PC, Myers RH, Feldman RG, Pollen D, Drachmann D, Growdon J, Bruni A, Foncin J-F, Salmon D, Frommelt P, Amaducci L, Sorbi S, Piacentini S, Stewart GD, Hobbs WJ, Conneally PM, Gusella JF (1987) The genetic defect causing familial Alzheimer's disease maps on chromosome 21 Science 235 885–890

St George-Hyslop P, McLachlan DC, Tsuda T, Rogaev E, Karlinsky H, Lippa CF, Pollen D (1994) Alzheimer's disease and possible gene interaction [letter] Science 263 537

Strittmatter WJ, Saunders AM, Schmechel D, Pericak-Vance M, Enghild J, Salvesen GS, Roses AD (1993a) Apolipoprotein E High avidity binding to β-amyloid and increased frequency of type 4 allele in late-onset familial Alzheimer disease Proc Natl Acad Sci USA 90 1977–1981

Strittmatter WJ, Weisgraber KH, Huang DY, Dong L-M, Salvesen GS, Pericak-Vance M, Schmechel D, Saunders AM, Goldgaber D, Roses AD (1993b) Binding of human apolipoprotein E to synthetic amyloid β peptide. isoform-specific effects and implications for late-onset Alzheimer disease Proc Natl Acad Sci USA 90. 8098–8102

Strittmatter WJ, Saunders AM, Goedert M, Weisgraber KH, Dong L-M, Jakes R, Huang DY, Pericak-Vance M, Schmechel D, Roses AD (1994a) Isoform-specific interactions of apolipoprotein E with microtubule-associated protein tau implications for Alzheimer disease. Proc Natl Acad Sci USA 91 11183–11186

Strittmatter WJ, Weisgraber KH, Goedert M, Saunders AM, Huang D, Corder EH, Dong L-M, Jakes R, Alberts MJ, Gilbert JR, Han S-H, Hulette C, Einstein G, Schmechel DE, Pericak-Vance MA, Roses AD (1994b) Hypothesis Microtubule instability and paired helical filament formation in the Alzheimer disease brain are related to apolipoprotein E genotype Exp Neurol 125: 163–171

Tanzi RE, Gusella JF, Watkins PC, Bruns GAP, St George-Hyslop P, Van Keuren ML, Patterson D, Pagan S, Kurnit DM, Neve RL (1987) Amyloid beta protein gene cDNA, mRNA distribution and genetic linkage near the Alzheimer locus Science 235. 880–884

Tsuda T, Lopez R, Rogaeva EA, Freedman M, Rogaev E. Drachman D, Pollen D, Haines J, Liang Y, Crapper McLachlan DR, Duara R, St George-Hyslop P (1994) Are the associations between Alzheimer's disease and polymorphisms in the apolipoprotein E and the apolipoprotein CII genes due to linkage disequilibrium? Ann Neurol 36 97–100

Van Broeckhoven C, Backhovens H, Cruts M, Martin JJ, Crook R, Houlden H, Hardy J (1994) APOE genotype does not modulate age of onset in families with chromosome 14 encoded Alzheimer's disease. Neurosci Lett 169 179–180

Wisniewski T, Castano EM, Golabek A, Vogel T, Frangione B (1994) Acceleration of Alzheimer's fibril formation by apolipoprotein E in vitro Am J Pathol 145 1030–1035

Yu C-E, Payami H, Olson JM, Boehnke M, Wijsman EM, Orr HT, Kukull WA, Goddard KAB, Nemens E, White JA, Alonso ME, Taylor TD, Ball MJ, Kaye J, Morris J, Chui H, Sadovnick AD, Martin GD, Larson EB, Heston LL, Bird TD, Schellenberg GD (1994) The apolipoprotein E/CI/CII gene cluster and late-onset Alzheimer disease Am J Human Genet 54 631–642

Role of Apolipoprotein E in Alzheimer's Disease: Clues from its Structure

*K.H. Weisgraber** and *L.M. Dong*

Introduction

A major advance toward understanding the etiology of Alzheimer's disease (AD) was the discovery by Duke University investigators that one of the common alleles of the plasma lipid-transporting protein, apolipoprotein (apo) E, was a major susceptibility gene (Strittmatter et al. 1993a; Corder et al. 1993; Saunders et al. 1993; Roses et al. 1995). It was also demonstrated that allelic variation at the apoE gene locus affects the progression of the disease in an isoform-specific and dose-dependent manner: apoE4 > apoE3 > apoE2 (Roses et al. 1995; Corder et al. 1994). This breakthrough focused attention on and brought into clearer perspective a series of observations suggesting that apoE was linked to neurobiology in addition to its long-known role as a transporter of plasma lipids and modulator of plasma lipoprotein triglyceride and cholesterol metabolism (for reviews, see Mahley 1988 and Weisgraber 1994). Early studies had connected apoE to nerve regeneration following injury, where apoE participated in local capture and reutilization of neuronal membrane lipids at the injury site during the degeneration and regeneration processes (Boyles et al. 1989; Pitas et al. 1987; Skene and Shooter 1983; Müller et al. 1986; Ignatius et al. 1986). These studies led to the observation that the presence of apoE influences neurite outgrowth in primary neuronal cultures (Handelmann et al. 1992). More recently, apoE was shown to influence this outgrowth in an isoform-specific manner, with apoE3, the most common isoform, promoting neurite outgrowth and apoE4, the product of the AD susceptibility gene, retarding outgrowth (Nathan et al. 1994). This difference may have a profound effect on normal nerve maintenance and repair of neurons that occur during remodeling of synaptic connections.

Apolipoprotein E has been detected in two pathological markers of AD, amyloid plaques (Namba et al. 1991; Wisniewski and Frangione 1992) and neurofibrillary tangles (Crowther 1993). The genetic link of the apoE4 gene to AD led to a series of observations that provided two biochemical correlations with the genetic evidence. Of potential importance is the demonstration that

* Gladstone Institute of Cardiovascular Disease, Department of Pathology, University of California, San Francisco, CA 94144, USA

A D Roses et al (Eds)
Apolipoprotein E and Alzheimer's Disease

apoE interacts in an isoform-specific manner with the Aβ peptide (Strittmatter et al. 1993b) and tau (Strittmatter et al. 1994a,b), the two major components of plaques and tangles, respectively. Future understanding of the role of apoE in AD is likely to be built on previous links of apoE with neurobiology in the context of the genetic linkage studies. In addition, the wealth of information from over 20 years of research on the role of apoE in lipid and lipoprotein metabolism is likely to be invaluable. Undoubtedly, understanding the effect of apoE allelic variation on AD will require correlating apoE structure with known functions and potential neuronal interactions or, put another way, the functional consequences of structure variation.

Apolipoprotein E Structure

Apolipoprotein E contains 299 amino acids (M_r, 34000) in a single polypeptide chain (Strittmatter et al. 1994b). Genetic heterogeneity arises from the presence of three common alleles at a single apoE gene locus on chromosome 19. The three common isoforms, designated apoE2, apoE3, and apoE4, differ from each other by cysteine-arginine interchanges at positions 112 and 158. At these polymorphic sites, apoE3, the most common isoform, contains cysteine and arginine, respectively, and apoE4 contains arginine and apoE2 contains cysteine at both positions (Weisgraber et al. 1981). The genetic heterogeneity results in six common genotypes/phenotypes: three homozygous (apoE2/2, apoE3/3, and apoE4/4) and three heterozygous (apoE2/3, apoE2/4, and apoE3/4). Allelic heterogeneity appears to be absent in the apoE gene from other species. Of the nine other species that have been sequenced, seven have arginine at a position equivalent to position 112 in human apoE and two have cysteine (Weisgraber 1994). Thus, it was logical to assume that the arginine-containing species were "apoE4-like" and the cysteine-containing species were "apoE3-like." As will be discussed below, the situation is more complex.

Physical-biochemical studies established that apoE contains two independently folded structural domains: an amino-terminal domain (residues 1–191, 22-kDa fragment) and a carboxyl-terminal domain (residues 216–299, 10-kDa fragment; Fig. 1; Wettereau et al. 1988; Aggerbeck et al. 1988). The three-dimensional structure of the amino-terminal domain has been determined by x-ray crystallography (Wilson et al. 1991). It contains five α-helices, four of which are arranged in an anti-parallel four-helix bundle (Fig. 1). The detailed structure of the carboxyl terminus is unknown. It is depicted in Fig. 1 as containing primarily α-helices, which is consistent with circular dichroism measurements. The structural domains of apoE also define functional domains. The amino-terminal domain contains the region that binds to lipoprotein receptors (residues 136–150), and the carboxyl-terminal domain contains the major lipid binding elements of apoE. Detailed studies examining the lipoprotein-binding properties of a series of carboxyl-terminal truncations have localized the major elements involved in lipoprotein association in the

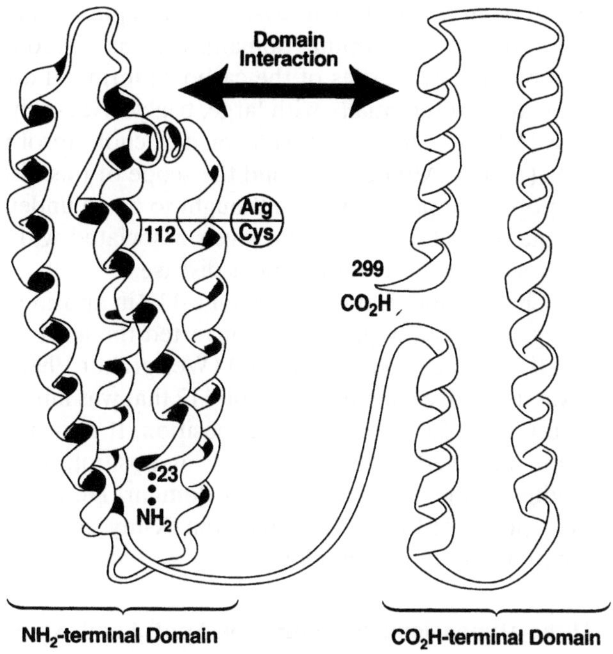

Fig. 1. Model of apoE domain structure. The amino-terminal domain (residues 1–191) contains the receptor-binding region of apoE (residues 136–150) and exists as a four-helix bundle The structure of the carboxyl-terminal domain is not known, but is depicted here as helical; it contains the major lipoprotein-binding determinants of apoE Position 112, which distinguishes apoE3 (Cys) from apoE4 (Arg), is indicated

carboxyl terminus to the vicinity of residues 244–272 and demonstrated that the extreme carboxyl terminus of the protein is not required for lipoprotein binding (Westerlund and Weisgraber 1993; Dong et al. 1994).

There is a significant impact of the cysteine–arginine interchange at position 158, which distinguishes apoE3 and apoE4 from apoE2. Whereas apoE3 and apoE4 with arginine at position 158 bind with high affinity to the low density lipoprotein (LDL) receptor, apoE2 with cysteine at the same position binds with low affinity. This defective binding impairs apoE-mediated lipoprotein clearance and is an underlying cause for the lipoprotein disorder, type III hyperlipoproteinemia (for review, see Mahley and Rall 1995).

One interesting and important aspect of apoE domain structure relates to the cysteine–arginine interchange at position 112, which distinguishes apoE3 and apoE4. A number of observations indicated that amino acid changes in one domain, or the domain itself, influenced the properties of the other domain (for review, see Weisgraber 1994). This concept is known as domain interaction. Much is known regarding the consequences of apoE3 and apoE4 domain interaction relative to lipoprotein metabolism. This information provides a valuable foundation for exploring apoE3 and apoE4 differences as they might relate to AD. The cysteine–arginine difference at position 112 lies within the amino-terminal domain, yet this amino acid difference affects the lipid binding properties of the carboxyl-terminal domain. As a result, apoE4 associates preferentially with large, triglyceride-rich very low density lipoproteins (VLDL) and apoE3 with smaller, cholesterol-rich high density lipoproteins (HDL). Although beyond the scope of this discussion, these differences in lipoprotein preference are thought to be an underlying cause for the higher plasma and LDL cholesterol levels associated with apoE4 and the resultant increased risk for cardiovascular disease.

Charge modification of cysteine-112 in apoE3 with cysteamine, a reagent that generates a lysyl analog, converted the apoE3 preference from HDL to VLDL, identical to that of apoE4 (Weisgraber 1990). This result indicated that it was a positive charge at position 112 that was important rather than a specific requirement for arginine at this position. It also suggested that domain interaction involved a positive charge at position 112, and probably an acidic residue in the carboxyl-terminal domain. Detailed structures of the apoE3 and apoE4 22-kDa fragments from X-ray crystallography showed that the situation was more complex.

Three-Dimensional Structures of ApoE3 and ApoE4

As shown Fig. 2, the 22-kDa fragments of both apoE3 and apoE4 fold into four-helix bundles, and with the exception of the residues that are shown, all the sidechains are essentially in identical conformations. In apoE3, cysteine-112 is located in a pocket between helices 2 and 3 and glutamic acid-109 occupies the position indicated. Arginine-61 lies over the pocket or groove formed by helices 2 and 3 (Wilson et al. 1991). However in apoE4 with arginine-112 in this pocket, glutamic acid-109 now forms a salt bridge with arginine-112, and the arginine-61 sidechain is forced away from the bundle and into the aqueous environment (Dong et al. 1994). This situation suggested that either the availability of glutamic acid-109 in apoE3 or the shift in the arginine-61 sidechain in apoE4 was responsible for mediating the interaction with the carboxyl-terminal domain. Site-directed mutagenesis of these residues distinguished between the two possibilities. Mutation of glutamic acid-109 to alanine in apoE3 did not affect the HDL preference. In contrast, mutating arginine-61 to threonine in apoE4 converted the lipoprotein preference from VLDL to HDL, thus establishing that arginine-61 was the key residue in the amino-terminal domain of apoE4 involved in domain interaction (Dong et al. 1994).

A series of carboxyl-terminal truncations of apoE3 and apoE4 demonstrated that residues 260–272 contained the major determinants for VLDL binding and the lipoprotein binding preferences, that residues 1–272 mimicked intact apoE, and that residues 273–299 were not essential for these characteristics (Dong et al. 1994). Mutation of each of the three acid residues between positions 260 and 272 did not affect the VLDL preference of apoE4,

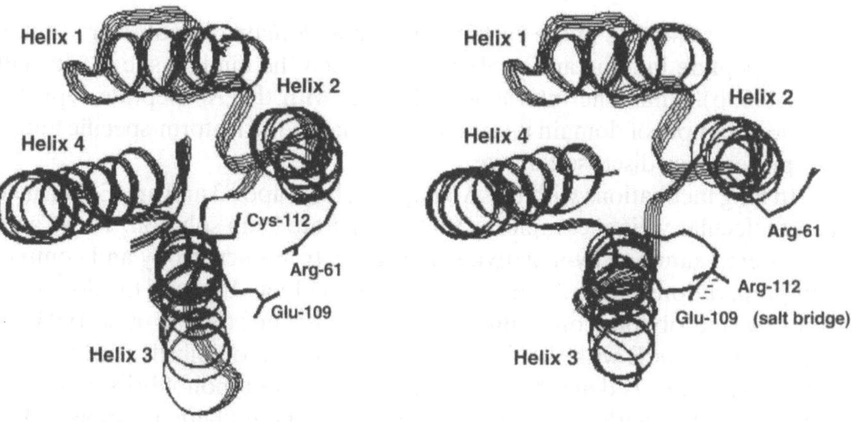

Fig. 2. Comparison of the X-ray crystal structures of the amino-terminal domain fragments of apoE3 and apoE4 Ribbon diagram of the four-helix bundle of the apoE3 (*left*) and apoE4 (*right*) 22-kDa fragments indicating the sidechain differences in the vicinity of position 112, which distinguishes apoE3 from apoE4

indicating that the interacting acidic residue(s) was more toward the amino terminus. The next three acidic residues in the vicinity of this region were mutated. Positions 244 and 245 were without effect. However, mutation of glutamic acid-255 shifted the preference of apoE4 from VLDL to HDL resulting in a distribution essentially identical to that of apoE3. Thus, glutamic acid-255 is the acidic residue within the carboxyl terminus that interacts with arginine-61 within the amino terminus to direct binding of apoE4 to VLDL. In the absence of this interaction, as occurs with apoE3, the default binding is to HDL.

Although the apoE from most other species (nine are known) contains arginine at a position equivalent to 112 in human apoE and might be construed as being "apoE4-like," the other species without exception lack the equivalent arginine-61 (Weisgraber 1994). Threonine occupies this position in all other species, ranging from monkeys to mice and rabbits. Thus, human apoE with its arginine-61 is unique among the species. This difference may have profound consequences relative to AD, as will be discussed below.

Isoform-Specific Interactions in Alzheimer's Disease

ApoE Interaction with the Aβ Peptide

The Aβ peptide and apoE interact in an isoform-specific manner and provided the first potential biochemical correlation with the genetic evidence consistent with a detrimental effect of apoE4 in AD. It was demonstrated that apoE4 binds more effectively to the Aβ peptide and that the pH range of interaction was different for the two isoforms (Strittmatter et al. 1993b). It also was shown that residues 244–272 in the carboxyl-terminal domain of apoE were required for Aβ peptide binding and probably contained the binding site (Strittmatter et al. 1993b). Thus, the interaction of apoE with the Aβ peptide represents another example of domain interaction, similar to the isoform-specific lipoprotein preferences discussed above.

In long incubations with the Aβ peptide, both apoE3 and apoE4 form very high molecular weight complexes that precipitate from solution. These aggregates were examined by negative staining electron microscopy and compared with preparations of the Aβ peptide incubated alone. By itself the Aβ peptide forms twisted ribbons containing two or more strands (Sanan et al. 1994). Coincubation of apoE with the Aβ peptide produced monofibrils approximately 7 nm in diameter. A denser, more extensive matrix of monofibrils was formed with apoE4 than with apoE3. Furthermore, the monofilaments appeared earlier with apoE4 than with apoE3. With both isoforms, immunogold labeling demonstrated that apoE was distributed along the entire length of the monofilaments. These results demonstrate that both isoforms interact with the Aβ peptide to form novel monofibrillar structures, with apoE4 being more effective. These isoform-specific effects with respect to fibril formation have been confirmed (Ma et al. 1994; Wisniewski et al. 1995).

It is reasonable to speculate that Aβ peptide produced by neurons or other cells may form monofibrils more readily in the presence of apoE4 and could potentially promote amyloid deposits. It is interesting that a similar region in apoE interacts with the Aβ peptide and is involved in lipoprotein binding and the isoform-specific lipoprotein preferences. This parallel suggests that the interaction of arginine-61 and glutamic acid-255 also influences the interaction apoE4 with the Aβ peptide.

ApoE Interaction with tau

In contrast to the isoform-specific interaction of apoE4 with the Aβ peptide, apoE3 forms a sodium dodecyl sulfate (SDS)–stable complex with tau, a member of the microtubule binding protein family. Little, if any, complex is formed between tau and apoE4 (Strittmatter et al. 1994b). The interaction of apoE3 with tau has been hypothesized to account for the apparent protective nature of apoE3 relative to apoE4 (for a review, see Strittmatter et al. 1994a). The binding of apoE3 with tau could stabilize microtubules and the cytoskeleton and perhaps help maintain the structure and function of neurons. Phosphorylated tau does not complex with apoE3. The binding of apoE3 with tau might also inhibit phosphorylation of tau and thus retard the paired–helical filament formation involved in the development of neurofibrillary tangles. Because apoE4 does not have the protective features of apoE3, these possibilities are consistent with the hypothesis that apoE4 is associated with AD.

It is critical to understand the molecular details of the apoE3–tau interaction and its implications on microtubule and cytoskeleton stability. In contrast to the interaction of the Aβ peptide with the carboxyl-terminal domain of apoE, the amino-terminal domain contains the tau-binding site (Strittmatter et al. 1994b). This suggests that the binding site is probably in the vicinity of the polymorphic site at position 112 and may lie in the grove between helices 2 and 3 (Fig. 2). This location implicates cysteine/arginine-112, arginine-61, and glutamic acid–109 as key residues in determining the specificity of the interaction.

Conclusion

The genetic link of apoE with AD has opened new and potentially productive areas of investigative research. Of particular importance are the biochemical correlations that have emerged involving apoE isoform-specific effects on neuronal outgrowth and isoform-specific interactions with the Aβ peptide and tau. Structural differences in the apoE isoforms and the functional consequences of these differences are key to delineating the role of apoE in AD. Understanding the isoform-specific interactions at the molecular level is likely to provide the basis for future therapeutic intervention and treatment of AD. The uniqueness of human apoE with respect to arginine-61 may be of particular importance since it is an important structural discriminator between apoE3 and apoE4. This raises the tantalizing question: Is AD limited to humans and could the absence of arginine-61 in the apoE of other species account for the difficulty in developing an animal model for AD?

Acknowledgments. This work is supported in part by the National Heart, Lung, and Blood Institute Program Project Grant HL 41633. The authors thank Kerry Humphrey for manuscript preparation, Amy Corder and John Carroll for graphics, and Dawn Levy and Gary Howard for editorial assistance.

References

Aggerbeck LP, Wetterau JR, Weisgraber KH, Wu C-SC, Lindgren FT (1988) Human apolipoprotein E3 in aqueous solution II Properties of the amino- and carboxyl-terminal domains. J Biol Chem 263. 6249–6258

Boyles JK, Zoellner CD, Anderson LJ, Kosik LM, Pitas RE, Weisgraber KH, Hui DY, Mahley RW, Gebicke-Haerter PJ, Ignatius MJ, Shooter EM (1989) A role for apolipoprotein E, apolipoprotein A-I, and low density lipoprotein receptors in cholesterol transport during regeneration and remyelination of the rat sciatic nerve J Clin Invest 83. 1015–1031

Corder EH, Saunders AM, Strittmatter WJ, Schmechel DE, Gaskell PC, Small GW, Roses AD, Haines JL, Pericak-Vance MA (1993) Gene dose of apolipoprotein E type 4 allele and the risk of Alzheimer's disease in late onset families. Science 261 921–923

Corder EH, Saunders AM, Risch NJ, Strittmatter WJ, Schmechel DE, Gaskell PC Jr, Rimmler JB, Locke PA, Conneally PM, Schmader KE, Small GW, Roses AD, Haines JL, Pericak-Vance MA (1994) Protective effect of apolipoprotein E type 2 allele for late onset Alzheimer disease. Nat Genet 7· 180–184

Crowther RA (1993) Tau protein and paired helical filaments of Alzheimer's disease Curr Opin Struct Biol 3 202–206

Dong L-M, Wilson C, Wardell MR, Simmons T, Mahley RW, Weisgraber KH, Agard DA (1994) Human apolipoprotein E Role of arginine 61 in mediating the lipoprotein preferences of the E3 and E4 isoforms J Biol Chem 269 22358–22365

Ignatius MJ, Gebicke-Harter PJ, Skene JHP, Schilling JW, Weisgraber KH, Mahley RW, Shooter EM (1986) Expression of apolipoprotein E during nerve degeneration and regeneration Proc Natl Acad Sci USA 83 1125–1129

Handelmann GE, Boyles JK, Weisgraber KH, Mahley RW, Pitas RE (1992) Effects of apolipoprotein E, β-very low density lipoproteins, and cholesterol on the extension of neurites by rabbit dorsal root ganglion neurons in vitro J Lipid Res 33 1677–1688

Ma J, Yee A, Brewer HB Jr, Das S, Potter H (1994) Amyloid-associated proteins α_1 -antichymotrypsin and apolipoprotein E promote assembly of Alzheimer β-protein into filaments. Nature 372 92–94

Mahley RW (1988) Apolipoprotein E cholesterol transport protein with expanding role in cell biology. Science 240 622–630

Mahley RW, Rall SC Jr (1995) Type III hyperlipoproteinemia (dysbetalipoproteinemia): the role of apolipoprotein E in normal and abnormal lipoprotein metabolism In Scriver CR, Beaudet AL, Sly WS, Valle D (eds) The metabolic and molecular bases of inherited disease Ed 7 McGraw-Hill, New York, 1953–1980

Muller HW, Ignatius MJ, Hangen DH, Shooter EM (1986) Expression of specific sheath cell proteins during peripheral nerve growth and regeneration in mammals J Cell Biol 102 393–402

Namba Y, Tomonaga M, Kawasaki H, Otomo E, Ikeda K (1991) Apolipoprotein E immunoreactivity in cerebral amyloid deposits and neurofibrillary tangles in Alzheimer's disease and kuru plaque amyloid in Creutzfeldt–Jakob disease Brain Res 541 163–166

Nathan BP, Bellosta S, Sanan DA, Weisgraber KH, Mahley RW, Pitas RE (1994) Differential effects of apolipoproteins E3 and E4 on neuronal growth in vitro. Science 264 850–852

Pitas RE, Boyles JK, Lee SH, Foss D, Mahley RW (1987) Astrocytes synthesize apolipoprotein E and metabolize apolipoprotein E-containing lipoproteins Biochim Biophys Acta 917 148–161

Roses AD, Saunders AM, Corder EH, Pericak-Vance MA, Han S-H, Einstein G, Hulette C, Schmechel DE, Holsti M, Huang D, Risch NJ, Haines JL, Goedert M, Jakes R, Dong L-M, Weisgraber KH, Strittmatter WJ (1995) Influence of the susceptibility genes apolipoprotein E-ε4 and apolipoprotein E-ε2 on the rate of disease expressivity of late-onset Alzheimer's disease Arzneim-Forsch/Drug Res 45 413–417

Sanan DA, Weisgraber KH, Russell SJ, Mahley RW, Huang D, Saunders A, Schmechel D, Wisniewski T, Frangione B, Roses AD, Strittmatter WJ (1994) Apolipoprotein E associates with β amyloid peptide of Alzheimer's disease to form novel monofibrils Isoform apoE4 associates more efficiently than apoE3 J Clin Invest 94 860–869

Saunders AM, Strittmatter WJ, Schmechel D, St George-Hyslop PH, Pericak-Vance MA, Joo SH, Rosi BL, Gusella JF, Crapper-MacLachlan DR, Alberts MJ, Hulette C, Crain B, Goldgaber D, Roses AD (1993) Association of apolipoprotein E allele ε4 with late-onset familial and sporadic Alzheimer's disease Neurology 43 1467–1472

Skene JHP, Shooter EM (1983) Denervated sheath cells secrete a new protein after nerve injury Proc Natl Acad Sci USA 80 4169–4173

Strittmatter WJ, Saunders AM, Schmechel D, Pericak-Vance M, Enghild J, Salvesen GS, Roses AD (1993a) Apolipoprotein E high avidity binding to β-amyloid and increased frequency of type 4 allele in late-onset familial Alzheimer disease Proc Natl Acad Sci USA 90· 1977–1981

Strittmatter WJ, Weisgraber KH, Huang DY, Dong L-M, Salvesen GS, Pericak-Vance M, Schmechel D, Saunders AM, Goldgaber D, Roses AD (1993b) Binding of human apolipoprotein E to synthetic amyloid β peptide isoform-specific effects and implications for late-onset Alzheimer disease Proc Natl Acad Sci USA 90 8098–8102

Strittmatter WJ, Weisgraber KH, Goedert M, Saunders AM, Huang D, Corder EH, Dong L-M, Jakes R, Alberts MJ, Gilbert JR, Han S-H, Hulette C, Einstein G, Schmechel DE, Pericak-Vance MA, Roses AD (1994a) Hypothesis Microtubule instability and paired helical filament formation in the Alzheimer disease brain are related to apolipoprotein E genotype Exp Neurol 125: 163–171

Strittmatter WJ, Saunders AM, Goedert M, Weisgraber KH, Dong L-M, Jakes R, Huang DY, Pericak-Vance M, Schmechel D, Roses AD (1994b) Isoform-specific interactions of apolipoprotein E with microtubule-associated protein tau Implications for Alzheimer disease Proc Natl Acad Sci USA 91 11183–11186

Weisgraber KH, Rall SC Jr, Mahley RW (1981) Human E apoprotein heterogeneity Cysteine-arginine interchanges in the amino acid sequence of the apo-E isoforms J Biol Chem 256 9077–9083

Weisgraber KH (1990) Apolipoprotein E distribution among human plasma lipoproteins role of the cysteine-arginine interchange at residue 112 J Lipid Res 31 1503–1511

Weisgraber KH (1994) Apolipoprotein E structure-function relationships Adv Protein Chem 45 249–302

Westerlund JA, Weisgraber KH (1993) Discrete carboxyl-terminal segments of apolipoprotein E mediate lipoprotein association and protein oligomerization J Biol Chem 268 15745–15750

Wetterau JR, Aggerbeck LP, Rall SC Jr, Weisgraber KH (1988) Human apolipoprotein E3 in aqueous solution I Evidence for two structural domains J Biol Chem 263: 6240–6248

Wilson C, Wardell MR, Weisgraber KH, Mahley RW, Agard DA (1991) Three-dimensional structure of the LDL receptor-binding domain of human apolipoprotein E Science 252 1817–1822

Wisniewski T, Frangione B (1992) Apolipoprotein E a pathological chaperone protein in patients with cerebral and systemic amyloid Neurosci Lett 135· 235–238

Wisniewski T, Lalowski M, Golabek A, Vogel T, Frangione B (1995) Is Alzheimer's disease an apolipoprotein E amyloidosis? Lancet 345 956–958

Apolipoprotein E4, Cholinergic Integrity, Synaptic Plasticity and Alzheimer's Disease

J. Poirier, M.-C. Delisle, R. Quirion, I. Aubert, J. Rocheford, I. Rousse, S. Gracon, M. Farlow,* and *S. Gauthier*

Summary

Recent evidence indicates that apolipoprotein E (apoE) plays a central role in the hippocampal response to injury. The coordinated expression of apoE and its receptor, the apoE/apoB (LDL) receptor, appears to regulate the transport of cholesterol and phospholipids during the early and intermediate phases of the reinnervation process. During dendritic remodelling and synaptogenesis, neurons progressively repress the synthesis of cholesterol in favor of cholesterol internalization through the apoE/LDL receptor pathway. The discovery that the apolipoprotein E4 allele is strongly linked to both sporadic and familial late onset Alzheimer's disease (AD) raises the possibility that a dysfunction of the lipid transport system associated with compensatory sprouting and synaptic remodelling could be central to the AD process. The role of apoE in the central nervous system is particularly important in relation to the function of the cholinergic system, which relies to a certain extent on the integrity of phospholipid homeostasis in neurons. Recent evidence indicates that the apoE4 allele has a direct impact on cholinergic function and on cholinomimetic drug response in AD subjects.

Introduction

Apolipoprotein E (apoE) is a well-characterized lipophilic protein associated with plasma and cerebral spinal fluid (CSF) lipoproteins. ApoE is synthesized primarily by the liver, but also at other sites including brain, macrophages and adrenals (Elshourbagy et al. 1985). Furthermore, apoE is unique among apolipoproteins in that it has a special relevance to the central and peripheral nervous systems. It is a key determinant in the cellular recognition and internalization of cholesterol- and phospholipid-rich lipoproteins in the developing brain and in response to neuronal injury (Boyles et al. 1989; Poirier et al. 1991a,b, 1993a). It was shown to play a fundamental role in the central

*McGill Centre for Studies in Aging, Douglas Hospital Research Centre, 6875 LaSalle Blvd., Verdun, Quebec H4H 1R3, Canada

A.D. Roses et al. (Eds.)
Apolipoprotein E and Alzheimer's Disease

nervous system during hippocampal synaptic remodelling induced by entorhinal cortex lesions in the rat (Poirier et al. 1991a,b, 1993a).

ApoE is synthesized and secreted by astrocytes in the deafferented zone of the hippocampus following lesions of the entorhinal cortex (Poirier et al. 1991a). Shortly after the peak in apoE synthesis, a time-dependent, cell specific induction of the [^{125}I]apoE/apoB (low density lipoprotein; LDL) receptors in the granule cell layer of the dentate gyrus was shown to correlate closely with the cholinergic synaptogenesis resulting from the loss of entorhinal cortex projections (Poirier et al. 1993a). In contrast, hippocampal cholesterol synthesis fell by more than 60% during the early phase of the cholinergic reinnervation process, during peak synthesis of apoE in the deafferented zone (Poirier et al. 1993a).

These changes were interpreted as evidence of a role for apoE and the LDL receptor in the process of cholinergic compensatory synaptogenesis caused by the loss of entorhinal cortex neurons, a pathophysiological process also observed in Alzheimer's disease (AD; Hyman et al. 1984).

The mature form of apoE present in human plasma and CSF is a single glycosylated 37-kDa polypeptide containing 299 amino acids (Rall et al. 1982). ApoE is a constituent of several plasma lipoproteins (very low density lipoprotein, VLDL, and high density lipoprotein, HDL) and mediates the cellular uptake of lipid complexes through interaction with the apoB/apoE (LDL) receptor and distinct hepatic apoE receptors. Human apoE is encoded by a four-exon gene (3.6 Kb in length) on the long arm of chromosome 19. Three major isoforms of apoE (E4, E3 and E2), differing by a single unit of net charge, can be easily detected by isoelectrofocusing (Utermann et al. 1979). These isoforms are expressed from multiple alleles at a single apoE genetic locus (Zannis and Breslow 1981) on chromosome 19, giving rise to the three common homozygous phenotypes (E4/4, E3/3 and E2/2) and three common heterozygous phenotypes (E4/3, E4/2 and E3/2).

Recently, the frequency of the apoE4 allele was shown to be markedly increased in sporadic (Poirier et al. 1993b; Saunders et al. 1993; Noguchi et al. 1993) and late onset familial AD (Corder et al. 1993; Payami et al. 1993). More interestingly, a gene dosage effect was observed in both familial (Corder et al. 1993) and sporadic (Poirier et al. 1993b) cases (i.e., as age of onset increases, ε4 allele copy number decreases). Women, who are generally at a greater risk of developing AD, showed increased apoε4 allele frequency when compared to age-matched men (Poirier et al. 1993b).

Apolipoprotein E4 and β Amyloid

ApoE-like immunoreactivity has been found to be associated with amyloid in senile plaques, cerebral vessels and neurofibrillary tangles (Namba et al. 1991). In vitro, amyloid β binds to purified apoE. According to Strittmatter et al. (1993), apoE4 binds more avidly to amyloid than apoE3, whereas LaDu et al. (1994) found that apoE3 has the highest affinity for the amyloid fragment. Increased amyloid β peptide deposition was recently reported in the cerebral cortex (Schmechel et al. 1993) and hippocampus (Beffert and Poirier 1995) of late-onset AD subjects carrying one or two copies of the apoE4 allele. Recently, it was shown that the interaction of synthetic amyloid with reconstituted apoE2, E3 and E4 can alter the binding properties of apoE toward its main receptor, the LDL receptor (Guillaume et al. 1995).

Apolipoprotein E4 and Cholinergic Dysfunction in Alzheimer's disease

Brain membrane phospholipids, especially phosphatidylcholine (PC) and phosphatidyl ethanolamine (PE), have been shown to play important roles in the availability of choline, a rate-limiting precursor of acetylcholine (ACh; Nitsch et al. 1992). The release from PC of free choline precursor for ACh synthesis is accomplished in a one-step process through a phospholipase-D type enzyme in cholinergic neurons. Brain levels of choline are decreased by up to 40%–50% in the frontal and parietal cortices (Nitsch et al. 1992) of AD patients. Similarly, cholesterol, which is decreased in the same brain areas in AD, is apparently required for the proper functioning of certain cholinergic receptor sub-types. As losses of cholinergic neurons and/or ChAT activity are well-known, classical hallmarks of AD (Etienne et al. 1986; Perry et al. 1977; Davies and Moloney 1976), a possible relationship between the apoE4 genotype and cholinergic deficits was deemed highly relevant to investigate. A marked reduction in ChAT activity in the hippocampus and temporal cortex of AD cases was reported to be inversely proportional to the apoε4 allele copy number (i.e., as apoE4 allele copy number increased, ChAT activity decreased; Poirier et al. 1994; Poirier 1994; Soininen et al. 1995).

The importance of apoE in the CNS has been further highlighted by the work of Masliah and colleagues (1995), who recently demonstrated that homozygous apoE-deficient (knockout) mice display significant loss of synapses and a marked disruption of the dendritic cytoskeleton in neurons with age. But more importantly, they showed that the hippocampal area of knockout mice cannot induce compensatory synaptogenesis following removal of entorhinal cortex projections. Low concentrations of brain apoE have also been observed in apoE4 carriers when compared to normal control subjects and to AD apoE3/3 subjects (Bertrand et al. 1995). These studies clearly suggested that low levels of apoE in the CNS of rodents and humans could compromise lipid homeostasis, selectively affect the cholinergic system and impair cognitive processes in animal models as well as in humans.

To test this hypothesis, we examined the effect of a total lack of apoE (apoE knockout mice) on cognitive performance in the Morris swim maze test. We further extended the original observation of an apoE4-mediated loss of cholinergic function by examining another pre-synaptic marker of cholinergic activity, the nicotinic receptor. Finally, we examined the impact of apoE genotype on cholinomimetic response in living AD subjects.

Results and Discussion

On the basis of its well-recognized post-mortem stability, choline acetyltransferase (ChAT) activity was examined in relation to apoE genotype in control (n = 30) and AD (33) post-mortem subjects. Figure 1 illustrates that the loss in ChAT activity (as measured by the method of Tucek 1978) in the hippocampus and temporal cortex of control and AD cases is a function of apoε4 allele copy number (i.e., as apoE4 allele copy number increased, ChAT activity decreased; Poirier et al. 1994). A similar genotype-dependent reduction of the nicotinic binding sites was also observed in the hippocampus and cortex of apoE4/AD carriers.

Nicotinic receptor binding site density decreases in AD as a function of genotype in the order of E3/2 >> E3/3 > E4/3 = E4/4, which parallels with the known genotype-dependent reduction in apoE concentration in the serum and brain of humans (Davignon et al. 1988; Bertrand et al. 1995). These changes are also consistent with the fact that nicotinic receptors, like ChAT, have a preferred pre-synaptic location in the hippocampal formation. In contrast, pirenzepine sensitive "M1" sites remain relatively unaltered in the hippocampus of AD subjects with different apoE genotypes (not shown).

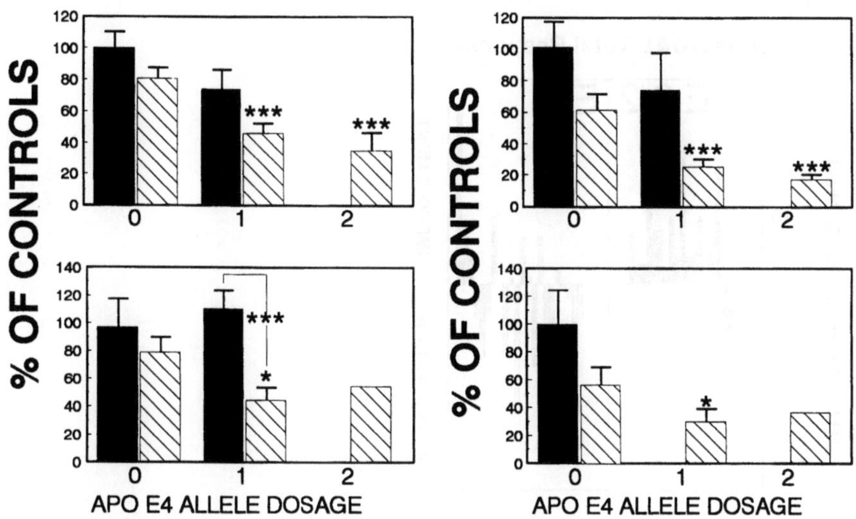

Fig. 1. Choline acetyltransferase activity (*top*) and density of nicotinic-binding sites (*bottom*) in the hippocampus (*left*) and temporal cortex (*right*) of control subjects (*black bars*) and Alzheimer's patients (*shaded bars*) using the method of Aubert et al (1992) Genotype was determined as described in Poirier et al (1994) The duration of disease was similar for both groups (AD-E3/3, 7 ± 3 years, AD-E4/4, 10 ± 4 years, mean ± S D.). Sex and age were matched as for as possible

We recently proposed that synaptic plasticity and cholinergic integrity may be selectively compromised in apoE4 carriers (Poirier 1994), as both processes rely on intact brain lipid homeostasis. Under these circumstances, drugs designed to take advantage of the residual cholinergic activity present in the AD brain should be more efficient in brains capable of plasticity and showing near normal ChAT and nicotinic receptor activities, i.e., in apoE3/E2 subjects. To test this hypothesis, the apoE genotype was determined in 40 patients enrolled in a 30-week randomized controlled trial of a high-dose of tacrine (Knapp et al. 1994), a potent, centrally active, cholinesterase inhibitor (Dawson 1989). Patients were selected on the basis of their drug responsiveness and completion of the trial (50% of subjects showing improvements after drug treatment and 50% showing deterioration). Cognitive performance (measured by the ADAS) was contrasted in AD subjects with and without the apoε4 allele.

Improvement was assessed using the standard Alzheimer's Disease Assessment Scale (Farlow et al. 1992; ADAS total and cognitive) at initial screen and after 30 weeks, at the conclusion of this trial.

Figure 2 illustrates individual resulting differential ADAS scores (ADAS end of trial − ADAS at screening) as a function of apoE genotype. Results indicate that 83% of AD subjects *not carrying* the apoε4 allele showed improvements in response to tacrine administration, whereas nearly 60% of apoE4 carriers showed significant deterioration with this treament. These

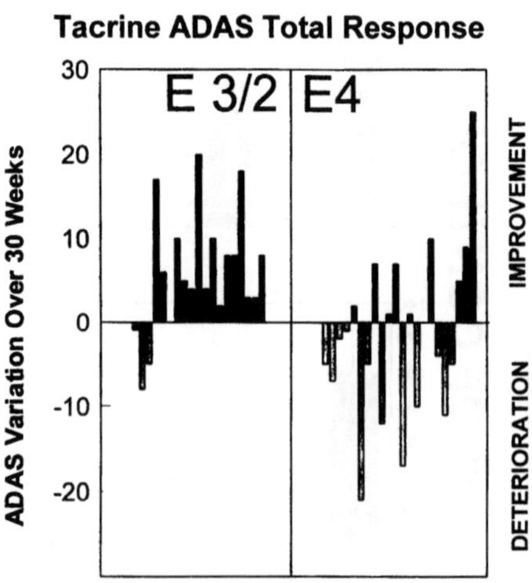

Fig. 2. Effect of apoε4 allele copy numbers on tacrine drug responsiveness in Alzheimer's disease (AD) subjects. Forty subjects enrolled in the 30-week randomized controlled trial of high-dose tacrine in patients with AD were selected from the original 663-subject cohort (Knapp et al 1994) Patients were selected prior to apoE phenotype determination and blind to genotype Phenotypic determination of the apoE genotype was performed as described before using frozen serum Graph represents individual variation (*ADAS* score before and after drug trial) in cognitive and non-cognitive performances as a function of apoE4 allele incidence

results are in good agreement with our initial working hypothesis that subjects not carrying the apoε4 allele should respond better to cholinergic drug therapies, with apoE3 and/or apoE2 bearers showing sufficient residual ChAT activity to benefit from an AchE-based therapy.

Finally, we examined the impact of a total loss of apoE on cognitive performance in young, 2-month-old apoE knockout mice using the Morris swim maze paradigm (Morris 1985). As cognitive performance (spatial memory and learning) in this test relies heavily (but not exclusively) on intact cholinergic functions, we hypothesize that the absence of apoE in these animals compromises cholinergic function and, indirectly, cognitive performance. Indeed, Fig. 3 displays the mean (±SEM) latency to find the platform for the apoE knockout and control mice across five days of testing. The latencies were analyzed by a strain X days analysis of variance, which yielded a significant interaction between these two factors ($p < 0.01$). Simple main effects F tests indicated that apoE knockout mice took significantly more time to find the platform on the first three days of testing, but not on the fourth. Performance on the visible platform trials also did not differ between the two strains. These results indicate that spatial learning and memory ability are impaired, but not effaced, in the apoE knockout mice.

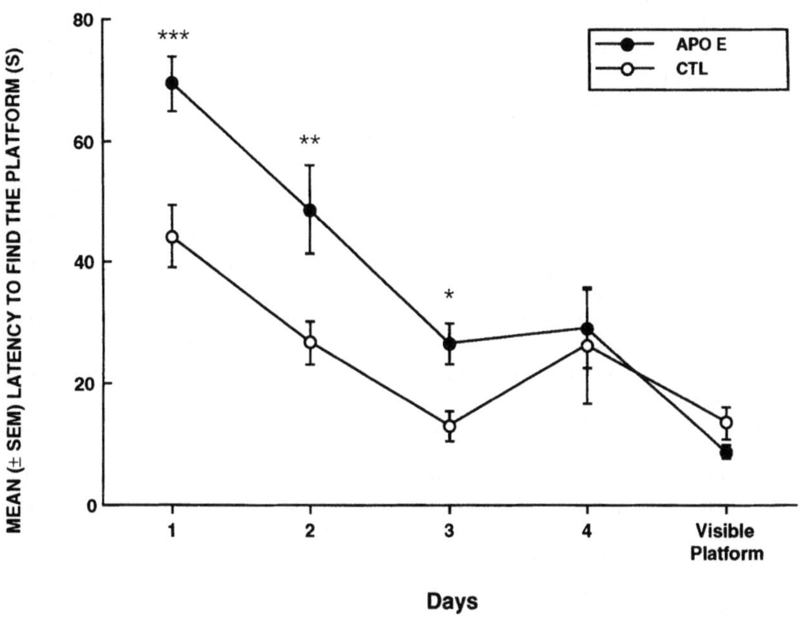

Fig. 3. Morris water maze latencies in apoE knockout mice as a measure of spatial learning and memory Ten animals were tested in each group (knockout and normal CTL C57bl mice) for 4 days. On the fifth day, the platform was made visible to identify possible visual deficits in the knockout mice *Asterisks* indicate significant differences between strains for each day: ***, $p < 0.001$, **, $p < 0.01$; *$p < 0.05$

Taken together, our data clearly suggest that apoE levels in the CNS may compromise cognitive performance in rodents and humans and that cholinergic function in AD-E3/3 subjects may at least be partially spared when compared to AD-E4/3 and AD-4/4 carriers. Similarly, mice lacking apoE demonstrate impaired cognitive performance when compared to mice expressing a normal concentration of apoE. Most importantly, the genetic susceptibility associated with the presence of apoE4 in humans resulted in distinct sub-groups of AD patients who responded differently to acetylcholine-based therapies (i.e., acetylcholine esterase inhibition), with E4 carriers losing more of their Ach synthetic capacities than E3 and E2 subjects.

Acknowledgments. The research from our laboratory reviewed here was supported by grants from the National Institute on Aging, AG-10003, le Fonds de la Recherche en Santé du Québec, Alzheimer Society of Canada and the Medical Research Council of Canada.

References

Aubert I, Araujo DM, Cécyre D, Robitaille Y, Quirion R (1992) Comparative alterations of nicotinic and muscarinic binding sites in Alzheimer's and Parkinson's diseases J Neurochem 58 529–541

Beffert U, Poirier J (1995) Apolipoprotein E, plaques, tangles and cholinergic dysfunction in Alzheimer's disease Ann NY Acad Sci, in press

Bertrand P, Oda T, Finch CE, Pasinetti GM, Poirier J (1995) Association between apolipoprotein E genotype with brain levels of apolipoprotein E and apolipoprotein J (clusterin) in Alzheimer's disease Mol Br Res, in press

Boyles JK, Zoellner CD, Anderson LJ, Kosick LM, Pitas RE, Weisgraber KH, Hui DY, Mahley RW, Gebicke-Haeter PJ, Ignatius MJ, Shooter EM (1989) A role for apolipoprotein E, apolipoprotein A-1, and low density lipoprotein receptors in cholesterol transport during regeneration and remyelination of the rat sciatic nerve J Clin Invest 83 1015–1031

Corder EH, Saunders AM, Strittmatter WJ, Schmechel DE, Gaskell PC, Small GW, Roses AD, Pericak-Vance MA (1993) Gene dose of apolipoprotein E type 4 and risk of Alzheimer's disease in late onset families Science 261 921–923

Davignon J, Gregg RE, Sing CF (1988) Apolipoprotein E polymorphism and atherosclerosis Arteriosclerosis 8. 1–21

Davies P, Maloney AJR (1976) Selective loss of central cholinergic neurons in Alzheimer's disease Lancet 2· 1403

Dawson RM (1989) Tacrine slows the rate of ageing of sarin-inhibited acetylcholine esterase Neurosci Lett 100 227–230

Elshourbagy NA, Liao WS, Mahley RW, Taylor JM (1985) Apolipoprotein E mRNA is abundant in the brain and adrenals, as well as in the liver, and is present in other peripheral tissues of rats and marmosets. Proc Natl Acad Sci 82 203–207

Etienne P, Robitaille Y, Wood P, Gauthier S, Nair NPV, Quirion R (1986) Nucleus basalis neuronal loss and choline acetyltransferase activity in advanced Alzheimer's disease Neuroscience 4 1279–1291

Farlow M, Gracon SI, Hershey LA, Lewis KW (1992) A controlled trial of tacrine in Alzheimer's disease JAMA 268 2523–2529

Guillaume D, Dea D, Davignon J, Poirier J (1995) Low density lipoprotein pathways in the central nervous system and apolipoprotein E isoform-specific differences In Iqbal K, Mortimer JA,

Winblad B, Wisniewski H (eds) Research advances in Alzheimer's disease and related disorders John Wiley and Sons, pp 384–395

Hyman BT, Van Hoesen GW, Damasio AR, Barnes CL (1984) Alzheimer's disease cell-specific pathology isolates the hippocampal formation. Science 225 1168–1170

Knapp MJ, Knopman DS, Solomon DR (1994) A 30-week randomized controlled clinical trial of high doses of tacrine in patients with Alzheimer's disease JAMA 271. 985–991

LaDu MJ, Falduto MT, Manelli AM, Reardon CA, Gertz GS (1994) Isoform-specific binding of apolipoprotein E to βamyloid J Biol Chem 269· 23403–23406

Masliah E, Mallory M, Alford M, Mucke L (1995) Abnormal synaptic regeneration in APP695 transgenic and apoE knockout mice In Iqbal K, Mortimer JA, Windblad B, Wisniewski HM (eds) Research advances in Alzheimer's disease and related disorders. Wiley J and Sons, 405–414

Morris RGM, Garrud P, Rawlings J, O'Keefe J (1982) Place navigation impaired in rats with hippocampal lesions Nature 297 681–683

Namba Y, Tomonaga M, Kawasaki H (1991) Apolipoprotein E immunoreactivity in cerebral amyloid deposits and neurofibrillary tangles in Alzheimer's disease and in Creutzfeldt-Jacob disease Brain Res 541 163–166

Nitsch RM, Blusztajn JK, Pitas AG, Slack BE, Wurtman PJ (1992) Evidence for a membrane defect in Alzheimer's disease Proc Natl Acad Sci USA 89 1671–1675

Noguchi S, Murakami K, Yamada N (1993) Apolipoprotein E and Alzheimer's disease Lancet (Letter) 342 737

Payami H, Kaye J, Heston LL, Schellenberg GD (1993) Apolipoprotein E and Alzheimer's disease Lancet (letter) 342. 738

Perry EK, Gibson PH, Blessed G (1977) Neurotransmitter enzyme abnormalities in senile dementia J Neurol Sci 34· 247–265

Poirier J (1994) Apolipoprotein E in animal models of brain injury and in Alzheimer's disease. Trends Neurosci 12 525–530

Poirier J, Hess M, May PC, Finch CE (1991a) Cloning of hippocampal poly(A+) RNA sequences that increase after entorhinal cortex lesion in adult rat Mol Brain Res 9: 191–195

Poirier J, Hess M, May PC, Finch CE (1991b) Apolipoprotein E and GFAP-RNA in hippocampus during reactive synaptogenesis and terminal proliferation. Mol Brain Res 11 97–106

Poirier J, Baccichet A, Dea D, Gauthier S (1993a) Role of hippocampal cholesterol synthesis and uptake during reactive synaptogenesis in adult rats Neuroscience 55 81–90

Poirier J, Davignon J, Bouthillier D, Bertrand P, Gauthier S (1993b) Apolipoprotein E and Alzheimer's disease Lancet 342 697–699

Poirier J, Aubert I, Bertrand P, Quirion R, Gauthier S, Nalbantoglu J (1994) Apolipoprotein E4 and cholinergic dysfunction in AD a role for the amyloid/apoE4 complex? In Giacobini E, Becker RE (eds) Alzheimer's disease therapeutic strategies Birkhauser, Boston, pp 72–76

Rall SC, Weisgraber KH, Mahley RW (1982) Abnormal lipoprotein receptor binding activity of the human apoE due to arginine-cysteine interchange at a single site J Biol Chem 257 4171–4178

Saunders AM, Strittmatter WJ, Schmechel D, St George-Hyslop PH, Pericak-Vance MA, Joo SH, Rosi BA, Gusella JF, McClaclans DR, Alberts MJ, Roses AD (1993) Association of apolipoprotein E allele E4 with late onset familial and sporadic Alzheimer's disease. Neurology 43 1467–1472

Schmechel D, Saunders AM, Strittmatter WJ, Crain BJ, Hulette CM, Joo SH, Roses AD (1993) Increased amyloid β-peptide deposition in cerebral cortex as a consequence of apolipoprotein E genotype in late-onset Alzheimer's disease Proc Natl Acad Sci USA 90 9649–9653

Soininen H, Kosunen O, Helisalmi S, Mannermaa A, Paljarvi L, Riekkinen O (1995) Severe loss of choline acetyltransferase in the frontal cortex of Alzheimer's disease patients carrying apolipoprotein E4 allele Neurosci Lett 187 79–82

Strittmatter WJ, Weisgraber KH, Huang DY, Dong LM, Salvesen GS, Pericak-Vance M, Schmechel D, Roses AD (1993) Binding of human apolipoprotein E to synthetic amyloid β-peptide isoform-specific effect and implication for late onset Alzheimer's disease Proc Natl Acad Sci USA 90 8098–8102

Tucek S (1978) Choline acetyltransferase In Acetylcholine synthesis in neurons London, Chapman and Hall Ltd , pp 29–42

Utermann G, Pruin N, Steinmetz A (1979) Apolipoprotein E polymorphism in health and disease Clin Genet 15· 37–62

Zannis VI, Breslow JL (1981) Human VLDL-apoE isoprotein polymorphism is explained by genetic variation and post-translational modifications Biochemistry 20 1033–1041

Pattern of Apolipoprotein E Immunoreactivity During Brain Aging

D.E. Schmechel, M.O. Tiller, P. Tong, M. McSwain, S.-H. Han, R. Ange, D.S. Burkhart,* and *M.K. Izard*

Abstract

Apolipoprotein E (APOE, gene; apoE, protein) ε4 allele is a risk factor for Alzheimer's disease (AD). Understanding the pathological basis for the effect of APOE alleles on disease onset is of key importance to rational treatment of AD. In this study, we employed specific immunocytochemistry to investigate the localization of apoE at different age points in several vertebrate species. These species included laboratory rat, normal and transgenic inbred mouse lines, several prosimian species, including *Galago* and *Otolemur*, and several baboons (*Papio papio*). Specificity of tissue fixation, processing and staining methods was confirmed by multiple controls including lack of staining of tissue from APOE knockout mice. Assessment of brain pathology included staining for Aβ-amyloid deposition, neurite formation, cholinergic dystrophy and glial cells.

ApoE-like immunoreactivity (apoE-LIR) is observed in the brains of all species, but with variable distribution. In rodents, apoE-LIR is generally observed in ependymal cells, choroid epithelial cells, microglial cells, and some fibrillary and protoplasmic astrocytes. Much of the staining in gray matter occurs in perineuronal glial nets. In primates, apoE-LIR is observed in the cell classes described above for rodents, but also in neurons even in young animals (10% maximum life span). In primates, most neuronal apoE-LIR is faint and perinuclear. In some brain regions, notably hippocampus, neuronal apoE-LIR is intense filling proximal and distal dendrites of pyramidal neurons. As for rodent brain, considerable staining is represented by perineuronal glial nets. In older *Otolemurs* with amyloid plaques, apoE-LIR stains the plaques in a manner indistinguishable from human cases. In brain regions with neuronal injury, apoE-LIR can be observed in neurites as well as in numerous microglia/macrophages.

Glial apoE-LIR is found in the white matter where glial secretion of apoE has been postulated to participate in synaptic remodeling, reverse cholesterol

*Departments of Medicine (Neurology) and Neurobiology, Joseph and Kathleen Bryan Alzheimer's Disease Research Center, P O. Box 2900, Duke University Medical Center, Durham, NC 27710, USA

A D Roses et al (Eds)
Apolipoprotein E and Alzheimer's Disease

transport, and tissue repair. In gray matter, glial apoE-LIR is also observed in perineuronal glial nets where release may also directly modulate the response of neurons to injury and age-related changes. The new finding is that intraneuronal apoE-LIR is present even in very young primates, and apparently represents a feature of normal development and aging. However, a role for intraneuronal apoE in abnormal aging is suggested by the presence of intensely apoE-LIR neurons in sectors of primate hippocampus implicated in human AD pathology. The presence of intracellular apoE in glial cells and in some sets of neurons as well as in extracellular deposits supports the concept that apoE is localized in multiple cellular and extracellular sites where isoform-specific effects on normal and abnormal human brain aging might occur.

Introduction

Apolipoprotein E (APOE, gene; apoE, protein) is a 34-kDalton protein coded for on human chromosome 19. The three major APOE alleles confer specific genetic susceptibility for the onset of Alzheimer's Disease (AD: Strittmatter et al. 1993a; Saunders et al. 1993a,b; Corder et al. 1993, 1994). Besides its known role in cholesterol transport in the periphery (Mahley 1988), apoE is also synthesized in the central nervous system by astrocytes and other glial derivatives, and participates in synaptic remodeling and repair and response to nervous system injury (Skene and Shooter 1983; Muller et al. 1985; Snipes et al. 1986; Ignatius 1986; Gelman et al. 1991; Handelmann et al. 1992; Poirier et al. 1991a,b, 1993, 1994). For a clearer understanding of how inheritance of the three human APOE alleles ($\varepsilon 2$, $\varepsilon 3$, and $\varepsilon 4$) might influence the onset of AD, it is important to define the timing and characteristics of apoE expression in relation to normal aging and to AD pathology.

First of all, apoE is localized in the intracellular and extracellular sites of AD pathology (Diedrich et al. 1991; Namba et al. 1991; Wisniewski and Frangione 1991; Strittmatter et al. 1993a; Schmechel et al. 1993; Rebeck et al. 1993; Han et al. 1994a,b). In our previous work, we demonstrated the presence of apoE immunoreactivity in senile plaques as well as in neurons with neurofibrillary tangles and in cortical neurons of an aged primate (Strittmatter et al. 1993a; Schmechel et al. 1993). In addition, we demonstrated the occurrence of apoE in some cortical neurons of hippocampus of control brains from *normal* nondemented persons dying of other causes, in some cortical neurons of temporal lobe specimens of young patients undergoing epilepsy surgery, as well as in neurons without neurofibrillary tangles in brains of AD patients (Han et al. 1994a,b). These results supported the notion that apoE might influence brain aging and neuronal pathology in aging and AD *at very early time points*. We have since gathered additional data on normal control autopsies with confirmation of our original results (Einstein 1995). The very key question of how and when APOE allele $\varepsilon 4$ affects disease onset of AD in humans would require careful analysis of pre-morbid cases of *all age ranges*. Such early time point cases, however, are the most difficult to obtain for studies of human brain aging and AD pathology, although the study of AD pathology in Downs syndrome has provided some information on this subject.

The examination of brain aging in animal models offers a number of advantages and disadvantages compared to studies of brain aging in human subjects. The major advantage of animal models is access to normal individuals at time points that are rarely available in human studies. Even when these early time points are available, they are usually represented by persons dying of disease. Another advantage of animal models is the ability to control or at least inventory genetic or environmental variability that may impact on brain aging.

In this study, we present information on the immunolocalization of apolipoprotein E in the brain tissue of several animal models of brain aging. We will emphasize the occurrence of neuronal localization of apoE and will employ the term apoE-like (apoE-LIR) immunoreactivity to denote the possibility of cross-reactivity of apoE epitopes. We have included information on rodent species (laboratory rat and normal and transgenic inbred mice strains) as well as several species of prosimian primates and selected primate samples. This material will be used to describe the relationship between specific patterns of apoE immunolocalization and other indices of brain aging such as gliosis, neuritic injury, and β-amyloidosis.

The major points to be made are: 1) there is considerable species variation in the pattern of apoE localization; and 2) localization of apoE-LIR is commonly observed in some cortical and subcortical neurons in primate species, varying from small amounts in many neuronal classes to strong staining of some cortical neurons in regions subject to aging-related changes; 3) intraneuronal apoE can be observed in both distal dendrites and axonal processes of certain cortical neurons; and 4) such intraneuronal apoE can be observed at very early time points in primate aging (10% of maximum life span). These results suggest that apoE has potentially "full access" to different neuronal compartments where it may interact with specific cellular proteins to influence cellular response to injury. Intraneuronal apoE may represent an additional and separate intracellular site in addition to apoE contained in glial cells and the well-described extracellular presence of secreted/deposited apoE. One or all of these anatomical sites may represent a location where human apoE alleles may influence the onset and outcome of age-related injury and AD pathology.

Case Material

Numerous cases were collected over a 15-year period from animals dying of natural causes, animals euthanized for non-treatable illness, and other healthy animals being utilized in scientific experiments where the animal was euthanized at the end of the experiment and where brain tissue was available. The animals utilized in the preparation of this report represent: 22 specimens of *Galago senegalensis (Moholi)*, aged newborn to 13 years old; 87 specimens of *Otolemur crassicaudatus*, aged newborn to 17.6 years old; 71 specimens of *Otolemur garnetti* aged newborn to 22 years old; three specimens of *Lemur fulvus collaris*, aged 12, 21, and 34 years old, six young adult baboons (*Papio papio*); adult Sprague-Dawley laboratory rats, and inbred mice strains including wild type C57/B16j and APOE knockout mice (courtesy Dr. Nobuyo Maeda, University of North Carolina). While under deep anesthesia, animals were euthanized and various tissues including brain rapidly dissected. Brain tissue was cut in slabs of 2–3mm and placed in 10% buffered formaldehyde (commercial solution) or 4% buffered paraformaldehyde (prepared) for one to two days, and then transferred to phosphate buffer with slight a trace of formaldehyde left for longterm storage. In most cases, tissue was sectioned on Oxford vibratome at 35–45 microns and reacted as free-floating sections for histological procedures. In some cases, tissue was sectioned on cryostat or sliding microtome after cryoprotection, or embedded in paraffin using routine protocols of the general pathology laboratories. Not all cases in the above *Otolemur* series were incorporated in the graphs, but the selected subset are representative.

Immunocytochemistry and Histology

Immunocytochemistry was performed using avidin-biotin-peroxidase complex (ABC) methods using standard kits (Vector Labs, Burlingame, CA) and the following antibodies; ferritin (Boehringer-Mannheim), βA4 (mouse monoclonal "clone 10D5" to Aβ 1–28 courtesy of Athena Neuroscience), which recognizes beta-pleated Aβ fragment, choline acetyltransferase or ChAT (rat monoclonal from Boehringer-Mannheim), glial fibrillary acidic protein or GFAP (Boehringer-Mannheim), anti-HLA-DR (mouse monoclonal courtesy of Dr. Ottolenghi), Alz50 (monoclonal, gift of Dr. Peter Davies), and goat antisera to human apoE (Calbiochem) using diaminobenzidine as a detection method. Controls included method controls of non-brain specific antibodies for assessing non-specific staining of antisera, secondary reagents or ABC complex. In most cases, methanol-peroxide pretreatment (10% methanol-3% hydrogen peroxide) was employed to decrease endogenous peroxidase activity since most tissue was fixed for short time periods. For apoE immunocytochemistry, the best results were achieved with a pre-treatment period of 30 minutes to one hour with autologous serum for blocking, use of tris rather than phosphate buffers, and omission of warm temperatures during primary antibody incubation. Controls for the specificity of apoE reagent included demonstration of single band under similar conditions on Western blotting of liver or brain homogenates, non-staining of apoE knockout tissues including brain and liver (Fig. 1), absorption of antibody staining using immunoprecipitation, and lack of staining of other antisera directed to non-neural antigens. For apoE immunocytochemistry, further intensification of immunoreaction was achieved by pre-treatment with 10% formic acid. At dilutions of 1:10000 of anti-apoE, no bands were observed on Western blots at molecular weights of related proteins with shared epitopes such as apolipoprotein AI, AIV or lactotransferrin. Staining of apoE is somewhat reminiscent of lactotransferrin (Kawamata 1993; Leveugle 1994).

Ancillary histological material was prepared using thioflavine staining, Nissl stains using cresyl violet or thionine, and routine hematoxylin-eosin staining. In addition, some material was reacted with *Vicia villosa* agglutinin (VVA) coupled to biotin (Vector Labs, Burlingame, CA) and binding of VVA to perineuronal glial nets visualized with ABC methods (Bruckner et al. 1993; Seeger et al. 1994).

Analysis

Sections from each case were examined by several observers to confirm findings selected for photographic illustration. For estimates of plaque density, cholinergic neuritic change or number of ferritin-positive glial cells, selected fields were drawn using *camera lucida* on a 0.25 mm^2 grid and relevant areas/points of interest were tabulated by point counting methods. Such drawings were done in triplicate and the average score was reported for a given region, antibody, and animal. Usual variance was 5–10%.

Apolipoprotein E Immunolocalization in Rodents

In rodent brains, immunocytochemistry for apoE revealed immunoreactivity of choroid plexus, ependymal cells, and glial cells with morphology of astrocytes and microglial cells. Bergmann glial cells (radial glial derivative) of cerebellum were not stained. Under usual staining methods, rodent neurons were not immunoreactive although extensive perineuronal nets could be visualized as illustrated below fpr primates (Fig. 8C,D). In very old mice and some adult rats, occasional neurons were observed with faint immunoreactivity similar to the faint neurons illustrated in Fig. 8A,B. Often, such apoE-immunoreactive neurons were surrounded by a much more intense apoE immunoreactive net.

The availability of APOE knockout animals afforded an excellent test of specificity of apoE immunoreactivity. Tissue sections from liver of apoE knockout animals were unstained except for background compared to sections from wild-type litter mates obtained from heterozygote breeding processed in parallel (Fig. 1). The specificity of staining was also observed in brain tissue of wild-type and APOE knockouts. These experiments with APOE knockout

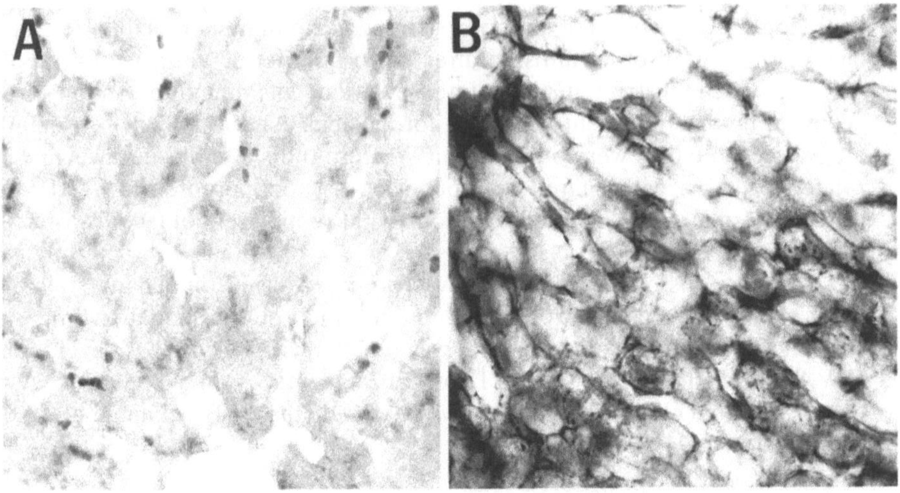

Fig. 1. Vibratome sections of mouse liver reacted for apoE immunolocalization showing lack of staining of liver from APOE knockout mouse and strong staining of liver from wild-type litter mate control (heterozygote breeding pairs) 320×

tissues confirmed the enhancement of apoE staining with formic acid pretreatments without loss of specificity, and led to the establishment of staining protocols described above.

Apolipoprotein E Immunolocalization in Primates

The oldest primate specimen available was a 34-year-old *Lemur fulvus collaris*, which is the animal reported on in our initial report on apoE (Strittmatter et al. 1993a). This animal had extensive Aβ-amyloidosis of cerebral vessels and had abundant cortical and striatal plaques. Using the commercially available Calbiochem goat antibody to apoE and revised protocols, we confirmed our previous results on neuronal localization of apoE, but obtained much better fidelity and strength of staining.

ApoE-like immunoreactive staining was observed in many pyramidal neurons throughout the cerebral cortex in this very old animal (near maximal life span in captivity for *Lemur fulvus collaris*). In the hippocampus and associated regions, the intensity of staining of cortical neurons in different cortical layers appeared laminar. For example, pyramidal neurons in superficial and deep layers of entorhinal cortex were more immunoreactive than in other layers of entorhinal cortex (Fig. 2A), although not as intensely stained as neurons in the CA1–2 and subiculum layers. In contrast, granule cell neurons in dentate gyrus were mostly unstained, although very light staining of some granule cell neurons contributed to vague definition of this layer (Fig. 2B). In subiculum and

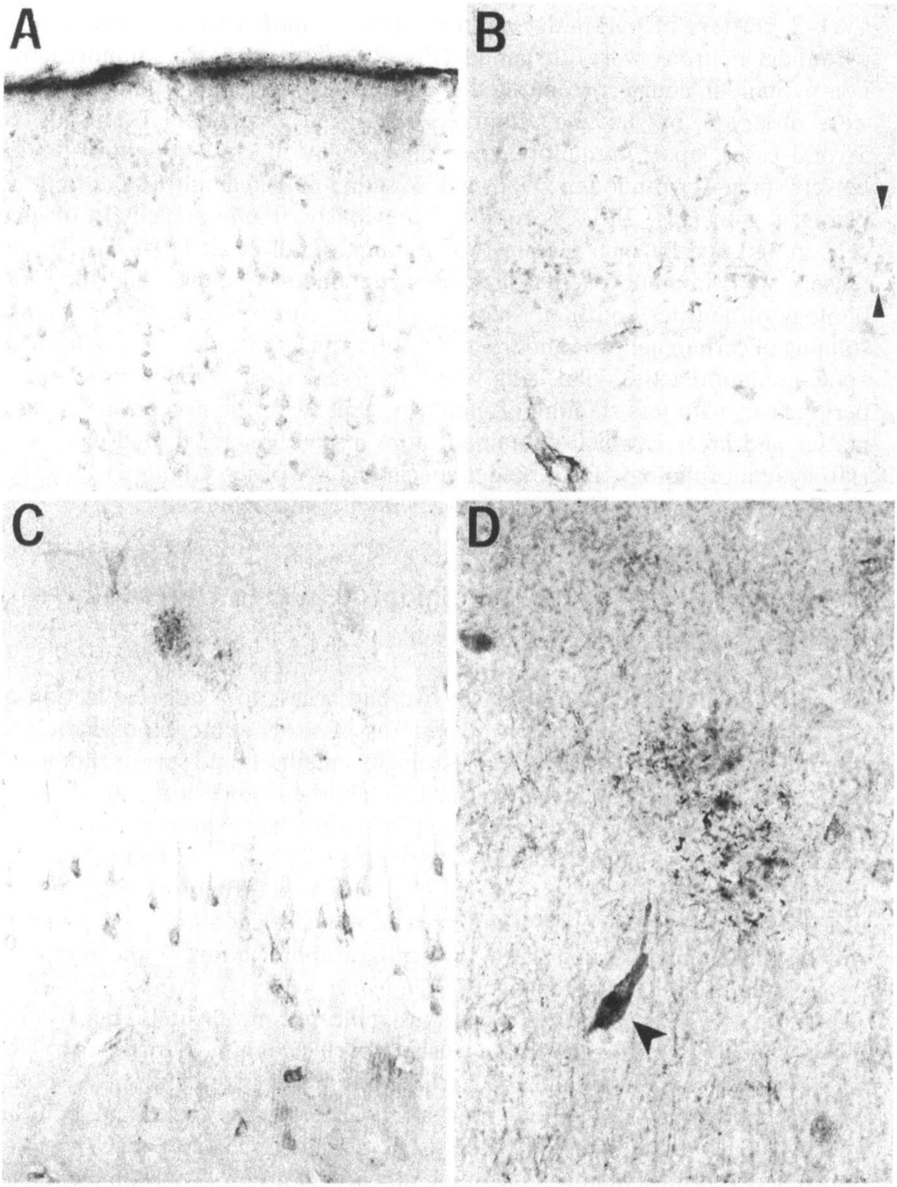

Fig. 2. Vibratome sections of brain of 38-year old *Lemur fulvus collaris* showing the presence of numerous APOE-like immunoreactive neurons in entorhinal cortex (**A**) In dentate gyrus, staining of neurons is even more heterogeneous with most granule cell neurons and basket neurons unstained while some few are faintly stained, often in clumps defining vaguely the outlines of the granule cell layer (extent indicated between two arrowheads) (**B**) In this animal, there were numerous Aβ-immunoreactive plaques and cerebral vessels that were also apoE-immunoreactive (see for example fragment of vessel in lower left of panel **B**) In the CA1-subiculum region of this animal, many pyramidal neurons are stained (**C**) Strongly apoE-immunoreactive neurons were often adjacent to plaques and their processes could be observed to contribute immunoreactive neurites to the plaque (**D**) 136×

CA1–2, clusters of pyramidal neurons showed apoE-LIR although nearby pyramidal neurons were unstained (Fig. 2C). Some apoE-immunoreactive non-pyramidal neurons (identified by dendritic morphology and location) were observed, but in lesser numbers compared to pyramidal neurons. In several cases, apoE-immunoreactive plaques lay in close proximity to intensely stained pyramidal neurons whose stained distal dendrites entered the plaque region (Fig. 2D). As reported previously, immunoreactivity of glial cells in and nearby such plaques was minimal (Han et al. 1994a,b). The intensely apoE-immunoreactive processes contained in plaques had the morphology of neurites and their staining intensity corresponded to the intense staining of certain neurons and their dendritic and axonal processes. Although apoE-immunoreactive glial cells were observed, their staining was mostly perinuclear, with less staining of glial processes, with the exception of white matter and layer I, which contained more intensely stained glial cells with astrocytic morphology. In the older specimens, we observed numerous small apoE-immunoreactive glial cells consistent with microglial cells.

Relationship of Neuronal Apolipoprotein E to Other Indices of Brain Aging

In the *Otolemur* and *Galago* species, we had access to a large collection of cases that permitted correlation of patterns of apoE immunolocalization to other indices of brain aging. Using histological methods and specific immunocytochemistry in this material, we have established numerical estimates for extent of cholinergic dystrophy, cell loss in the locus coeruleus and pendunculopontine-lateral dorsal tegmental (PPN-LDT) region of the brainstem, appearance of ferritin-rich glial cells, and formation of dystrophic processes (axonal spheroids) using markers for mitochondria. This material will be presented in graph form to facilitate appreciation of the temporal progression of brain aging changes in *Otolemur*.

In the *Otolemur* species in captivity, significant central cholinergic dystrophy occurs that can be recognized using specific immunocytochemistry for ChAT. The most prominent and earliest brain areas containing dystrophic central cholinergic projections that are those of the pedunculopontine nucleus-lateral dorsotegmental (PPN-LDT) cholinergic complex (CH5–6) to dorsal and ventral thalamus and other subcortical areas. Dystrophic endings are visible by two years of age in *Otolemur*, which represents approximately 10% of its maximum life span, and are most easily observed in the dense cholinergic innervation of the nucleus reticularis of thalamus (nRT) (Fig. 3). The apparent result of this age-related cholinergic dystrophy is progressive cholinergic denervation of the nRT (Fig. 4A,B), which is a linear process with time. Dystrophic serotonergic and noradrenergic endings are also observed with specific immunocytochemistry in these same animals (results not shown). Of note is the fact that the cholinergic injury process does not result in

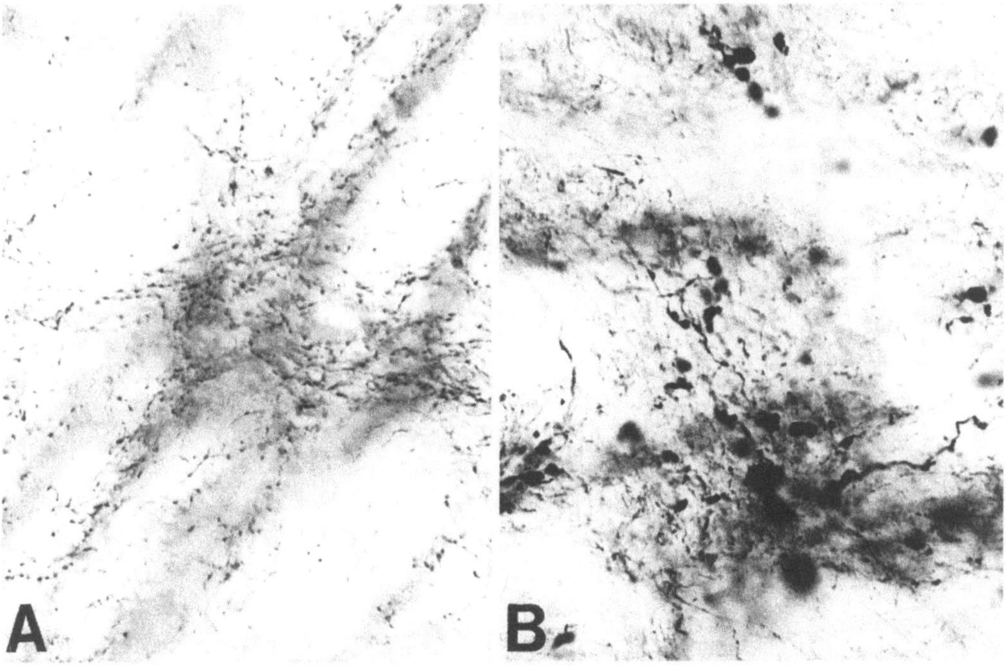

Fig. 3. ChAT immunocytochemistry of young *Otolemur* showing normal fine axonal morphology of cholinergic innervation of nucleus reticularis thalami (nRT) (**A**). Normal cholinergic axons appear as fine dots along reticular net of target neuronal cell bodies and dendrites of nRT In an older animal, significant numbers of misshapen and dystrophic cholinergic fibers are present (**B**), appearing as large heavy dots and strings over nRT. There is an associated loss of normal complement of fine, normal appearing cholinergic processes. 280×

apparent loss of neurons in the PPN-LDT complex (Fig. 4C) in contrast to the progressive age-related loss of noradrenergic neurons of the locus coeruleus (Fig. 4D). This process is much less severe in *Galago senegalensis* compared to *Otolemur* and thus does not seem to represent an inevitable consequence of brain aging in all primates.

The injury/aging process characterized by dystrophic neurites in *Otolemur* is accompanied by astrocytosis and microglial reaction that can be visualized by immunolocalization of ferritin, glial fibrillary acidic protein or by MHC class II antigens (Fig. 5). For example, even at the earliest time points, cholinergic dystrophy in the nRT is also associated with increasing numbers of ferritin-rich glia (Fig. 6A).

Dystrophic cholinergic fibers are also observed in hippocampus and other cortical areas in these same animals, although there are less numerous those found in thalamus. Dystrophy is not limited to central cholinergic fibers, being also observable for monoaminergic projections of the ascending reticular activating system using either specific immunocytochemistry or generalized methods of detecting injured nerve cell processes. Such a dystrophic process and associated spheroids from these and other central projections can be observed

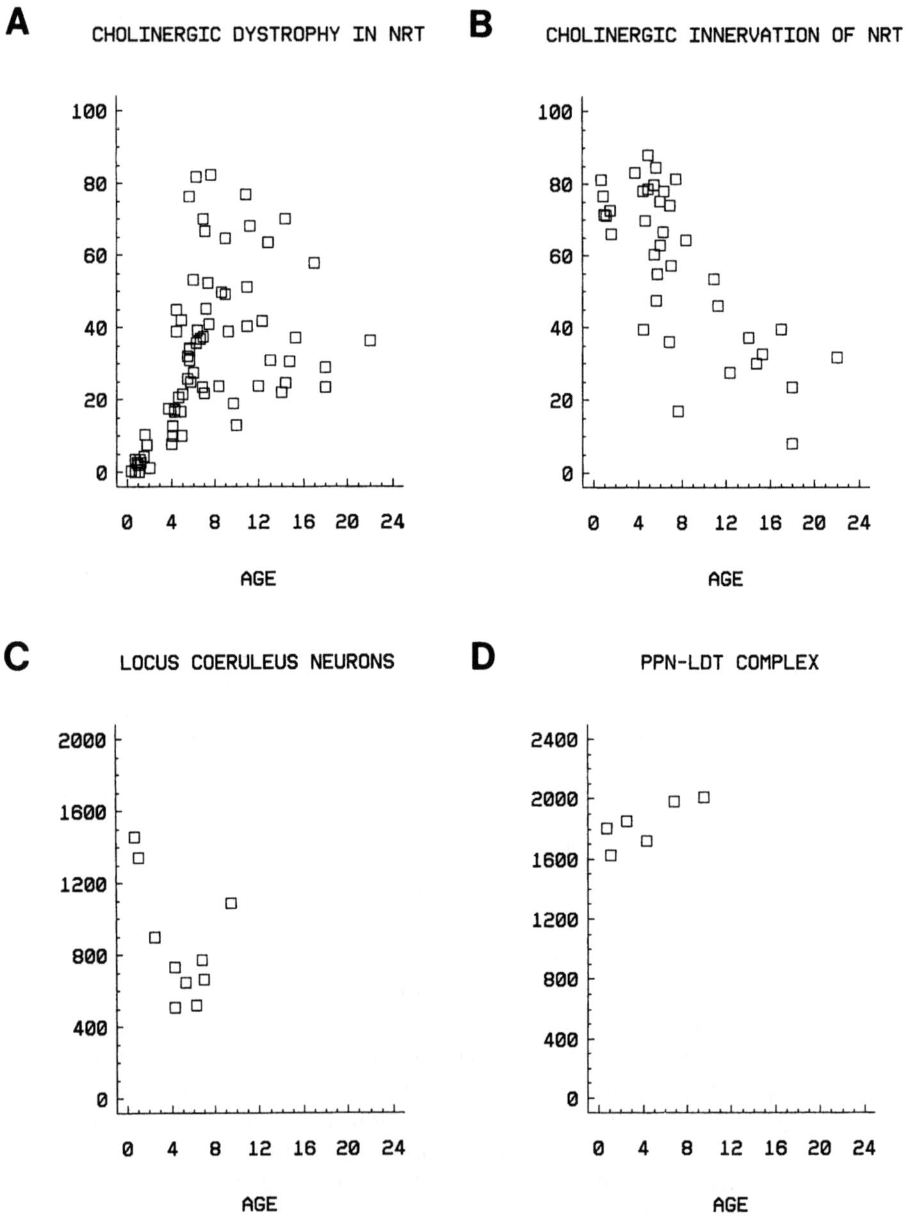

Fig. 4. Graphs comparing for *Otolemur* specimens of various ages the amount of cholinergic dystrophy in nRT (**A**), the associated cholinergic denervation of nRT (**B**), the loss of neurons in locus coeruleus (**C**), and the relative preservation of neurons in pedunculopontine-lateral dorsal tegmental nuclear complex (PPN-LDT) (**D**) Maximum life span in the wild is 7–10 years, and 20–22 years in captivity Ordinate is % area affected (0–100%) for panels **A** and **B**, and number of total cells for panels **C** and **D** Abscissa is age of animal in years

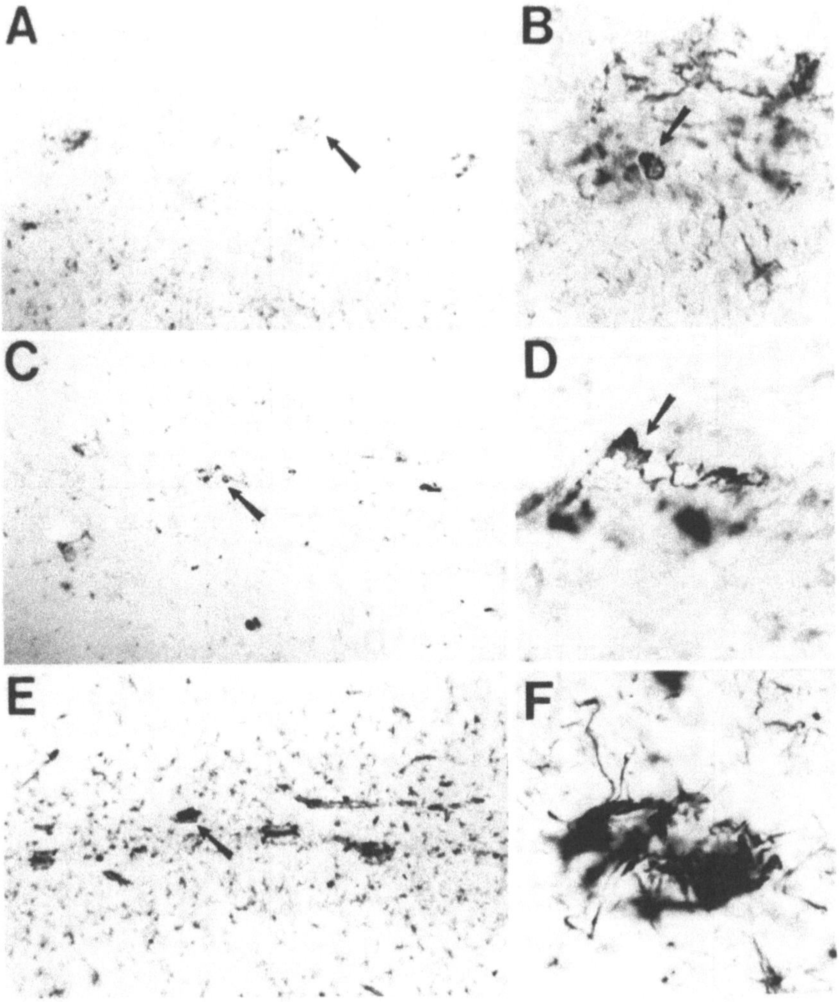

Fig. 5. Semi-adjacent sections of hippocampus of older *Otolemur* stained for immunolocalization of ferritin (**A** and **B**), HLA-DR for MHC II activation (**C** and **D**), and GFAP (**E** and **F**) This figure demonstrates the occurrence of widespread astrocytosis in aging *Otolemurs* which can be detected by increased ferritin content (all glial classes), of plaque-like foci of increased HLA-DR expression (presumably microglial cells), and by widespread astrocytosis (GFAP staining) Panels **A**, **C** and **D** 85×, and panels **B**, **D** and **F** show details of areas indicated by arrows at 600×

in layer I of neocortex and numerous other locales, most prominently the stratum lacunosum-moleculare and molecular layer of hippocampus in distribution of perforant pathway and in anterior lateral hypothalamus, nucleus accumbens and caudate nucleus (illustrated for hypothalamus in Fig. 6B).

Otolemurs kept in captivity and able to live long beyond their normal feral life span show aggressive brain aging as indexed by cholinergic dystrophy, loss

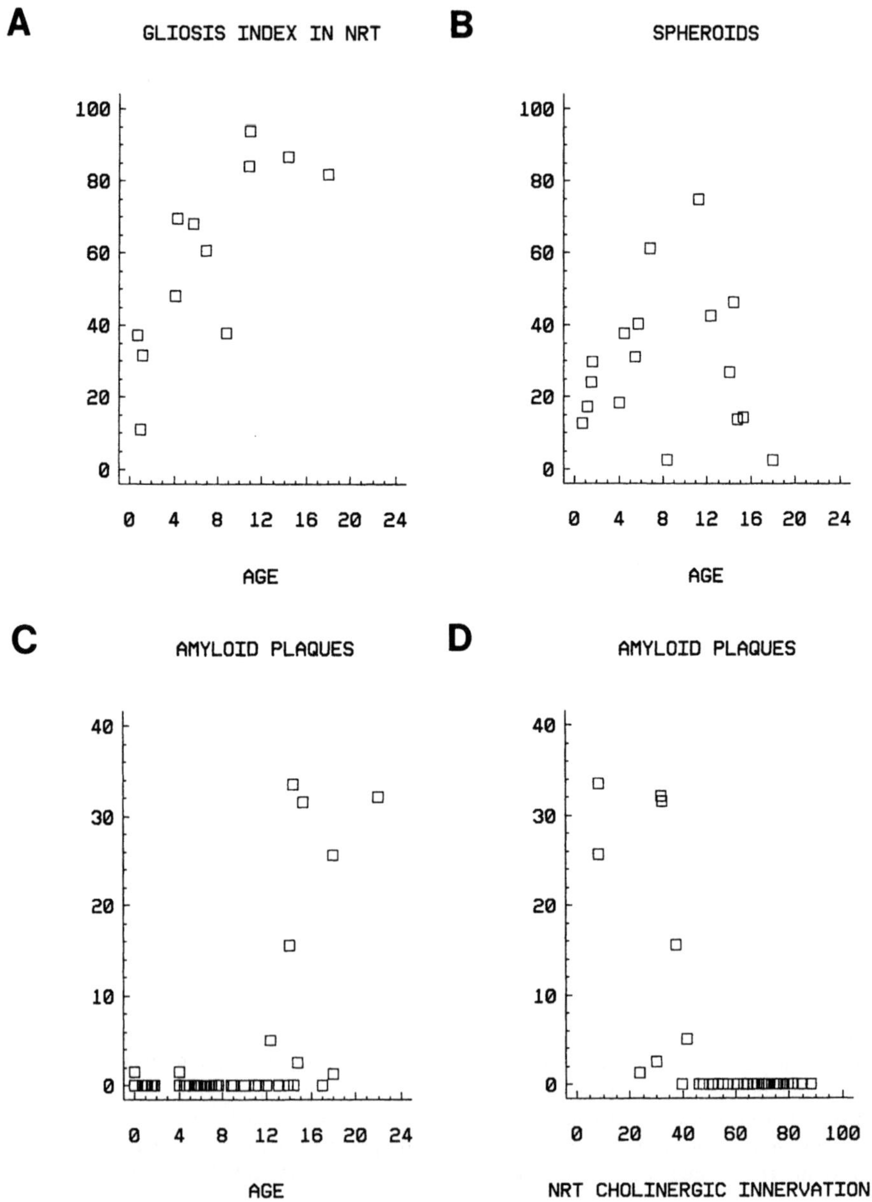

Fig. 6. Graphs depicting for *otolemur* specimens of various ages the amount of reactive gliosis as measured by intensely ferritin-immunoreactive glial profiles in nRT (**A**), and neuronal injury as measured by density of axonal spheroids in anterolateral hypothalamus (**B**) The occurrence of Aβ-amyloidosis as reflected by area density of Aβ-immunoreactive plaques or vessels is not linear, but appears first at roughtly 50% of maximum life span (**C**). When compared to the degree of cholinergic denervation of nRT, Aβ-amyloidosis is commonly observed in animals with greater than 60% loss of cholinergic innervation of nRT (**D**) Ordinate is % area affected (**A,B**) or covered (**C,D**) Abscissa is age in years

of locus coeruleus neurons, and gliosis. In fact, nearly all of these animals show extensive β-amyloidosis after 50% maximum life span with plaque densities approaching those of human cases of Alzheimers Disease by CERAD criteria (Mirra et al. 1991). The amount of Aβ-amyloidosis observed in brain aging of *Otolemurs* is indicated in panels C and D of Fig. 6, which shows the striking non-linear relationship to age. The distribution of Aβ-amyloidosis is most prominent in the limbic cortex, including the hippocampus, but does include visual cortex and caudate nucleus sparing most other subcortical areas including cerebellum. Extensive vascular Aβ-amyloidosis is also observed. Interestingly, the onset of Aβ-amyloidosis is closely correlated to the degree of cholinergic denervation of nRT, so that the one example of an older animal with minimal amyloidosis was also an animal with a lesser degree of cholinergic denervation (Fig. 6D).

In a subset of eight animals aged from two to 11 years, we explored the relationship of apoE immunolocalization to the above pattern of aging. We found that, even in the youngest animals, there were examples of intensely staining APOE neurons in all cortical regions. In Fig. 7, we illustrate an APOE-LIR pyramidal neuron in hippocampus (CA1-subiculum border) of a young *Otolemur*. This neuron is intensely stained compared to its neighbors, and demonstrates not only perinuclear staining but also staining of distal

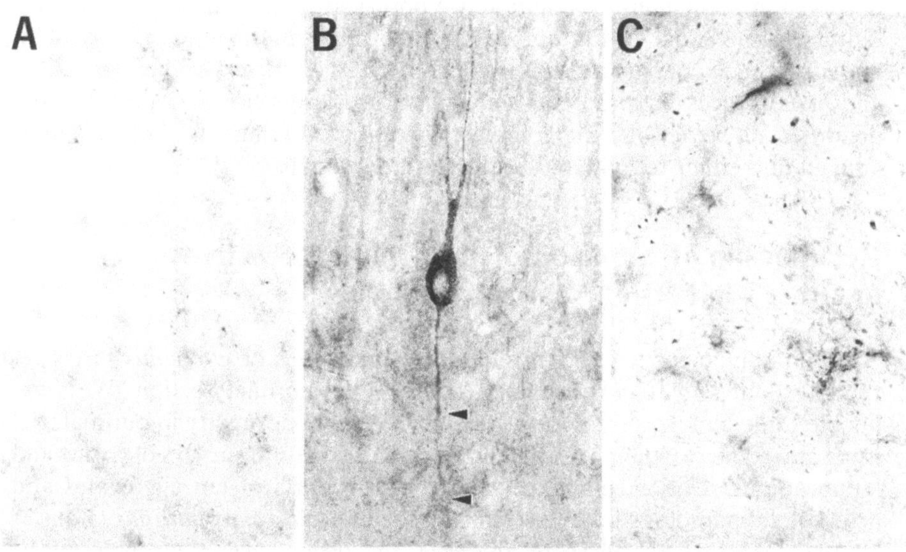

Fig. 7. ApoE-like neuronal immunoreactivity is observed in young adult *Otolemurs* In this example, a hippocampal pyramidal neuron is fully stained including secondary dendrites and axonal process (arrows) in (**B**), despite lack of staining of any nearby glial cells or neighboring neurons. An adjacent section reacted in parallel with control goat antiserum shows minimal background staining (**A**) Staining of scattered axons (small dark dots) is illustrated for subcortical white matter in aged lemur Some of specific staining in (**C**) compared to (**A**) is also attributable to immunostained fragments of glial cell processes 390×

dendrites to secondary branches, as well as staining of the axon. Staining of axons was also observed in the alveus and adjacent white matter to hippocampus of other older prosimians (Fig. 7C), although usually only a few axons stained intensely.

In this material, we could often observe neurons that were unstained or surrounded by apoE-immunoreactive nets (Fig. 8C,D) that may correspond to VVA-demonstrated perineuronal astrocytic nets (Bruckner et al. 1993; Seeger et al. 1994). Many other neurons had some amount of faint, perinuclear staining that was above background but not as intense (Fig. 8A,B). Indeed, many of these neurons had faintly stained nuclei. Finally, there were examples (5–15 cells per section) of more intensely stained pyramidal neurons even in the youngest specimen (Fig. 8). In the oldest *Otolemurs* of this subset (approximately 10 years of age), there were generally more strongly apoE-immunoreactive neurons (25–50 cells per section), but not to the extent observed in the even older animals (10–20 years of age) with significant Aβ-amyloidosis.

Our assessment of the primate material is that apoE-LIR of neurons is common, but may vary considerably in apparent amount and localization of apoE-LIR, ranging from none to low amounts, to in some instances large amounts filling both neuronal cell body and proximal/distal processes. In a separate set of apoE immunolocalization experiments in baboons, we also observed apoE immunolocalization in astrocytes and in neurons similar to that observed in prosimians. In addition, we observed specific apoE immunoreactivity of some axons in white matter even at high dilutions of antisera. This apoE-LIR of axonal processes is quite dependent on formic acid pretreatment and tissue processing protocols and subject to noise at times. ApoE in distal dendrites and axons of some neurons may represent relatively "hidden" apoE antigen due either to lipid and/or protein interactions.

Relationship of Apolipoprotein E-Immunoreactivity to Other Markers of Neuronal Pathology

Alz50 immunocytochemistry in the oldest specimens of *Otolemur* series revealed about two to three stained neurons per coronal section. We were unable to document any widespread cytoskeletal abnormality in our material using antibodies to human tau and limited ultrastructural surveys of cortex and hippocampus. This is contrasted with recent reports of tau aggregates and cytoskeletal pathology in the prosimian species *Microcebus murinus* (Bons et al. 1991; Bons et al. 1995). In any case, the occurrence of strongly apoE-immunoreactive neurons even in two year-old animals suggests that neurons may contain apoE long before any readily detectable cytoskeletal abnormalities supervene in some prosimian primates. This supports information from our laboratory from aged, non-demented human controls (Einstein 1995) and from temporal lobe specimens from persons undergoing surgery for complex partial seizures (Han et al. 1994a,b).

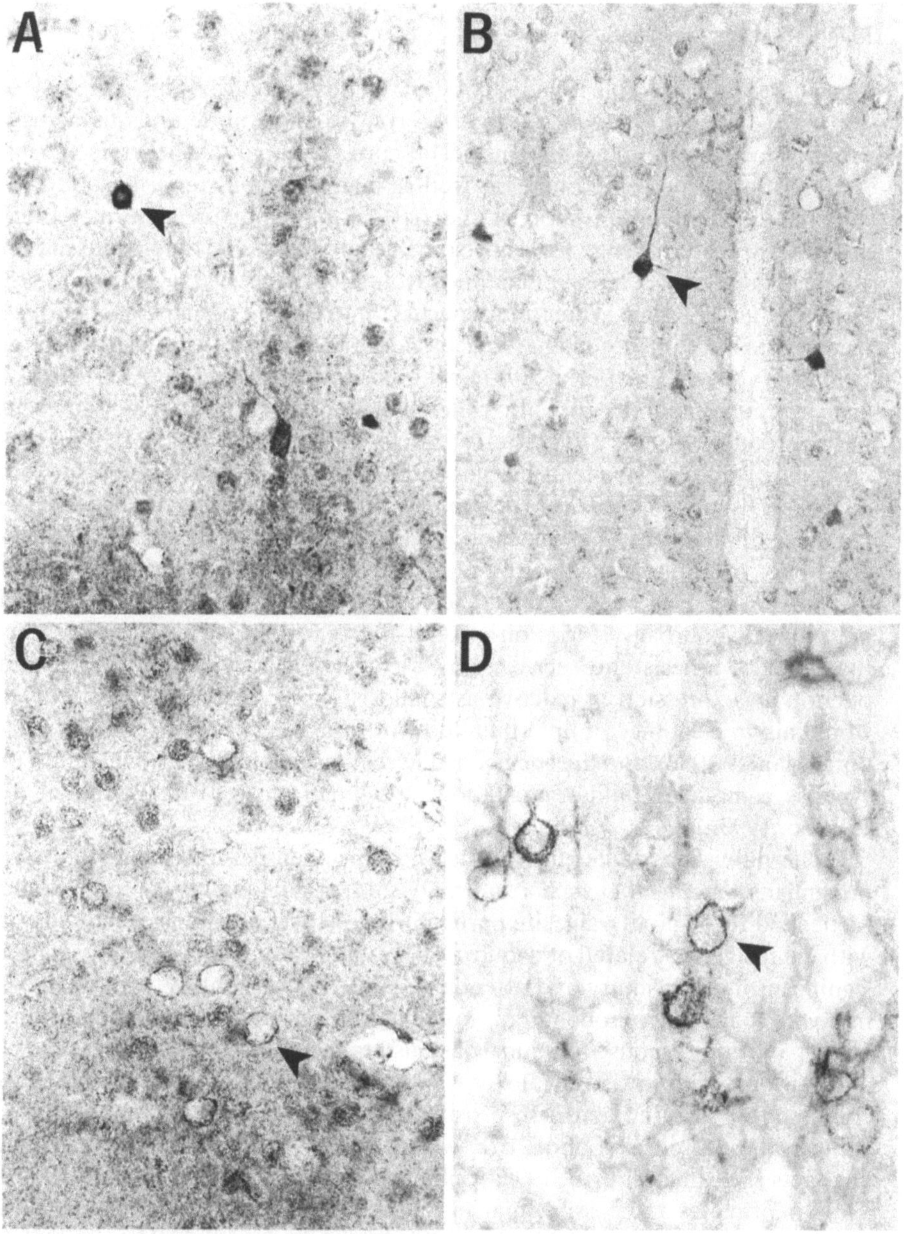

Fig. 8. Intensity and localization of apoE-like immunoreactivity of neurons varies considerably. In these examples from occipital lobe of a two-year-old *Otolemur*, scattered intensely stained cortical neurons can be observed (**A** and **B**), (arrowheads) with oftentimes considerable staining of apical and basal dendrites (**B**) Some neurons are non-immunoreactive Many neurons are faintly stained with granular deposits in the perinuclear area and even at times within the nucleus (**A** and **B**) In other instances, neurons are unstained, but surrounded by a dense, perineuronal net of apoE immunoreactivity (**C**) (see one example indicated by arrowhead) This appearance is quite similar to perineuronal glial nets that can be visualized by VVA lectin binding (**D**)

Discussion

Consideration of the role of human APOE alleles in contributing to the risk of AD must take into account the normal role of apoE in nervous tissue as a protein involved in injury response. The current immunocytochemical localization of apoE presented here, as well as many previous reports (see Introduction for references), support the notion that apoE is a normal brain protein, expressed at high levels by some astrocytes in gray and white matter and also present in cerebrospinal fluid (Roheim et al. 1979; Boyles et al. 1985; Poirier et al. 1991a,b). When injury occurs, apoE is one of many glial derived factors that are expressed in nervous tissue. As first illustrated by Diedrich for human brain (Diedrich et al. 1991), we observed abundant apoE in net-like arrays around neurons in both rodents and primates that may represent perineuronal glial nets (Bruckner et al. 1993; Seeger et al. 1994). Previous work mostly based on rodents has not emphasized the presence of apoE within neurons. Discussion of the role of apoE in injury response has centered on the response of glial cells (astrocytes) to neuronal injury and breakdown of synapses. For example, after deafferentation of rodent hippocampus, abundant mRNA for apoE is expressed in reactive astrocytes, but does not appear in neurons (Elshourbagy 1985; Poirier et al. 1991a,b, 1994). This work suggests that apoE synthesis and secretion by astrocytes is part of the genetic and phenotypic expression of reactive astrogliosis. ApoE secretion by astrocytes would aid in clearance of lipid breakdown through reverse cholesterol transport (Mahley 1988), and thus be needed in regions undergoing axonal injury or synapse remodeling and/or breakdown (Gelman et al. 1991; Poirier et al. 1991a,b, 1993, 1994).

The published studies in rodents and our extensive longitudinal data in prosimians suggests that such a "normal" role for glial-expressed apoE would be needed throughout adult life particularly for species, strains or individuals with extensive age-related or acquired CNS injury. For example, in *Otolemur*, significant neuritic injury and attendant gliosis are present by 10% of maximum life span. Unlike rodents, an explanation for primates and humans of the role of apoE in response to injury must also take into account the presence of apoE *within* neurons, even at very early time points in addition to glial-contained or secreted apoE. In prosimian, primate and human material, we confirm the presence of apoE-LIR within neurons and within perineuronal glial nets (Strittmatter et al. 1993a; Schmechel et al. 1993; Han et al. 1994a,b). Thus, in primates and possibly man, a role for apoE in normal brain aging and response to injury must consider a possible long-term positive role or longterm adverse role for intraneuronal apoE.

It is important to emphasize that the presence of intraneuronal apoE in *Otolemur* brain precedes the onset of $A\beta$-amyloidosis by 10 or more years, and is not accompanied by neruofibrillary tangles. Thus, concepts of aging-related and AD-related pathology might be better framed in the context of neuronal injury and associated glial reaction rather than a "cascade" of any particular protein, be it Aβ, apoE or some other protein. In *Otolemur*, the anatomically observable cascade that exists is the recruitment, reaction and persistence of glial cells in response to the neuronal injury. This glial "cascade" includes reactive astrocytes, microglial cells, and other elements which can then interact in a potentially synergistic manner for the better or worse of that particular region. The high iron content of perineuronal glial nets (Bruckner et al. 1993; Seeger et al. 1994) containing apoE might also promote oxidative change to apoE and other proteins in the near vicinity of the neuron.

The finding of apoE-like immunoreactivity in distal *dendrites and axon* of cortical pyramidal neurons of young animals is quite provocative with regard to theories of involvement of apoE alleles with specific neuronal cytoskeletal proteins (Strittmatter et al. 1993a,b, 1994a,b). In *Otolemur*, apoE would be able to interact with cellular organelles given apparent cytoplasmic localization (Han et al. 1994a,b) as well as with specific cytoskeletal proteins such as tau or MAP2C. These interactions would occur in selected sets of neurons for nearly the entire adult life.

ApoE is a difficult antigen in our experience, with possible lability, shared epitopes with other proteins, as well as potential lipid/protein interactions that may result in poor detection in immunocytochemical experiments. Two other groups have not reported apoE immunoreactivity in primate neurons (Poduri et al. 1994; Mufson et al. 1994). Our current evidence supports the presence of apoE in neurons of prosimian and primate species, and relative absence of apoE in neurons of normal rodent species. Our control experiments suggest that the immunochemical identity is apoE. Recognizing the possibility of cross-reacting or shared epitopes might account for some or all of the localization, we are currently employing the term apoE-like immunoreactivity as a cautionary reminder. On the other hand, it should be recognized that neurons have a number of receptors for internalization of apoE (e.g., Rebeck et al. 1993; Tooyama et al. 1993; Moestrup et al. 1992). One might well ask, therefore, why should rodent and primate neurons *not* contain detectable apoE given the presence of apoE receptors on neuronal cell bodies and the presence of apoE in perineuronal locations in glial processes and in the extracellular fluid?

Neuronally localized apoE in prosimians and humans may mark neurons that are responding to injury or involved in remodelling of synapses and membranes. It is important to state that we found different categories of apoE/neuronal content from no staining of neuron or surround, to staining of perineuronal glial nets, to light staining of perinuclear neuronal cytoplasm, to intensely stained neurons "filled" with apoE-like immunoreactivity. These neurons contain cytoplasmic apoE (Han et al. 1994b) similar to cytoplasmic localization of apoE described for hepatocytes (Hamilton et al. 1990).

Given the intense staining of some pyramidal neurons with apoE and their potential involvement in neurite formation, at least in the oldest animals, it will be important to determine the regulation of apoE uptake by neurons, the different compartments accessible to apoE, the possibility that iron-rich glial nets may provide a source of oxidized or otherwise damaged apoE to the nearby neuron, and lastly, the possibility that some neurons during response to injury may synthesize apoE themselves. These questions will require careful *in vitro* and *in vivo* experiments with improved reagents and better detection thresholds for demonstration of apoE protein and APOE mRNA.

Acknowledgments. This work was conducted with the support of NIA Alzheimer's Disease Center AG-05128, NIGMS Clinical Research Center grant RR-30, and numerous contributions, less than $100, to the Joseph and Kathleen Bryan Alzheimer's Disease Research Center. Additional support for portions of the work was NIEHS IAA Y01-ES-40290 for environmental factors in neurodegenerative diseases. We acknowledge the invaluable technical assistance of Susan Reeves for photography.

References

Bons N, Mestre N, Petter A (1991) Senile plaques and neurofibrillary changes in the brain of an aged lemurian primate, *Microcebus murinus* Neurobiol Aging 13· 99–105

Bons N, Jallageas V, Sihol S, Mestre-Frances N, Petter A, Delacourte A (1995) Immunocytochemical characterization of tau proteins during cerebral aging of the lemurian primate *Microcebus murinus*. CR Acad Sci Paris (Life Sci) 318 77–83

Boyles JK, Pitas RE, Wilson E, Mahley RW, Taylor JM (1985) Apolipoprotein E associated with astrocytic glia of the central nervous system and with nonmyelinating glia of the peripheral nervous system J Clin Invest 76 1501–1513

Bruckner G, Brauer K, Hartig W, Wolff JR, Rickmann MJ, Derouiche A, Delpech B, Girard N, Oertel WH, Reichenbach A (1993) Perineuronal nets provide a polyanionic, glia-associated form of microenvironment around certain neurons in many parts of the rat brain Glia 8 183–200

Corder EH, Saunders AM, Strittmatter WJ, Schmechel DE, Gaskell PC, Small GW, Roses AD, Haines JL, Pericak-Vance MA (1993) Gene dose of apolipoprotein E type 4 allele and the risk of Alzheimer's disease in late onset families Science 261: 921–923

Corder EH, Saunders AM, Risch NJ, Strittmatter WJ, Schmechel D, Gaskell P, Rimmler JB, Locke PA, Conneally PM, Schmader KE, Small GW, Roses AD, Haines JL, Pericak-Vance MA (1994) Protective effect of apolipoprotein E type 2 allele for late-onset Alzheimer disease. Nature Genet. 7 180–184

Diedrich JF, Minnigan H, Carp RI, Whitaker JN, Race R, Frey W II, Hasse AT (1991) Neuropathological changes in scrapie and Alzheimer's disease are associated with increased expression of apolipoprotein E and cathepsin D in astrocytes. J Virol 65 4759–4768

Elshourbagy NA, Liano WS, Mahley RW, Taylor JM (1985) Apolipoprotein E MRNA is abundant in the brain and adrenals, as well as in the liver, and is present in other peripheral tissues of rats and marmosets Proc Natl Acad Sci USA 82 203–207

Gelman BB, Goodrum JF, Bouldin TW (1991) Macrophage apolipoprotein synthesis and endoneurial distribution as a response to segmental demyelination J Neuropathol Exp Neurol 50 383–407

Hamilton RW, Wong JS, Guo LSS, Krisans S, Havel RJ (1990) Apolipoprotein E localization in rat hepatocytes by immunogold staining of cryothin sections J Lipid Res 31 1589–1603

Handelmann GE, Boyles JK, Weisgraber KH, Mahley RW, Pitas RE (1992) Effects of apolipoprotein E, β-very low density lipoproteins, and cholesterol on the extension of neurites by rabbit dorsal root ganglion neurons in vitro J Lipid Res 33 1677–1688

Han S-H, Hulette CM, Saunders AM, Einstein G, Pericak-Vance MA, Strittmatter WJ, Roses AD, Schmechel DE (1994a) Apolipoprotein E is present in hippocampal neurons without neurofibrillary tangles in Alzheimer's disease and in age-matched controls Exp Neurol 128: 13–26

Han S-H, Einstein G, Weisgraber KH, Strittmatter WJ, Saunders AM, Pericak-Vance M, Roses AD, Schmechel DE (1994b) Apolipoprotein E is localized to the cytoplasm of human cortical neurons a light and electron microscopic study J Neuropath Exp Neurol 53: 535–544

Ignatius MT, Gebicke-Harter PJ, Skene JHP, Shilling KH, Weisgraber KH, Mahley RW, Shooter EM (1986) Expression of apolipoprotein E during nerve degeneration and regeneration Proc Natl Acad Sci USA 83 1125–1129

Kawamata T, Tooyama I, Yamada T, Walker DG, McGeer PL (1993) Lactotransferrin immunocytochemistry in Alzheimer and normal human brain. Am J Pathol 142 1584–1585

Leveugle B, Spik G, Perl DP, Bouras C, Fillit HM, Hof PR (1994) The iron-binding protein lactotransferrin is present in the pathologic lesions in a variety of neurodegenerative disorders a comparative immunohistochemical analysis Brain Res 650 20–31

Lin C-T, Xu Y, Wu J-Y, Chan L (1986) Immunoreactive apolipoprotein E is a widely distributed cellular protein Immunohistochemical localization of apolipoprotein E in baboon tissues. J Clin Invest 78 947–958

Mahley RW (1988) Apolipoprotein E cholesterol transprotein with expanding role in cell biology Science 240 622–629

Mirra SS, Heyman A, McKeel S, Sumi SM, Crain BJ, Brownlee LM, Vogel FS, Hughes JP, van Belle G, Berg L (1991) The consortium to establish a registry for Alzheimer's disease (CERAD) Part II Standardization of the neuropathologic assessment of Alzheimer's disease Neurology 41 479–486

Moestrup SK, Gliemann J, Pallesen G (1992) Distribution of the α_2-macroglobulin receptor/low density lipoprotein receptor-related protein in human tissues Cell Tissue Res 269 375–382

Mufson EJ, Benzing WC, Cole GM, Wang H, Emerich DF, Sladek JR, Morrison JH, Kordower JH (1994) Apolipoprotein E-immunoreactivity in aged rhesus monkey cortex: colocalization with amyloid plaques Neurobiol Aging 15. 621–627

Muller HW, Gebick-Harter PJ, Hangen DH, Shooter EM (1985) A specific 37000-dalton protein that accumulates in regenerating but not in non-regenerating mammalian nerves Science 228 499–501

Namba Y, Tomonaga M, Kawasaki H, Otomo E, Ikeda K (1991) Apolipoprotein E immunoreactivity in cerebral amyloid deposits and neurofibrillary tangles in Alzheimer's disease and kuru plaque amyloid in Creutzfeldt-Jakob disease Brain Res 541: 163–166

Poduri A, Gearing M, Rebeck GW, Mirra SS, Tigges J, Hyman BT (1994) Apolipoprotein E4 and beta-amyloid in senile plaques and cerebral blood vessels of aged rhesus monkeys. Amer J Pathol 144 1183–1187

Poirier J (1994) Apolipoprotein E in animal models of CNS injury and in Alzheimer's disease TINS 17: 525–530

Poirier J, Hess M, May PC, Finch CE (1991a) Cloning of hippocampal poly(A) RNA sequences that increase after entorhinal cortex lesion in adult rat. Mol Brain Res 9 191–195

Poirier J, Hess M, May PC, Finch CE (1991b) Astrocytic apolipoprotein E mRNA and GFAP mRNA in hippocampus after entorhinal cortex lesioning. Mol Brain Res 11 97–106

Poirier J, Baccichet A, Dea D, Gauthier S (1993) Cholesterol synthesis and lipoprotein reuptake during synaptic remodelling in hippocampus in adult rats Neuroscience 55: 81–90

Rebeck GW, Reiter JS, Strickland DK, Hyman BT (1993) Apolipoprotein E in sporadic Alzheimer's disease allelic variation and receptor interactions. Neuron 11: 575–580

Roheim PS, Carey M, Forte T, Vega GL (1979) Apolipoproteins in human cerebrospinal fluid. Proc Natl Acad Sci USA 76: 4646–4649

Saunders AM, Strittmatter WJ, Schmechel D, St George-Hyslop PH, Pericak-Vance MA, Joo SH, Rosi BA, Gussella JF, Crapper-MacLachlan DR, Alberts MJ, Hulette C, Crain B,

Goldgaber D, Roses AD (1993a) Association of apolipoprotein E allele ε4 with late-onset familial and sporadic Alzheimer's disease Neurology 43 1467–1472

Saunders AM, Schmader K, Breitner JCS, Benson MD, Brown WT, Goldfarb L, Goldgaber D, Manwaring MG, Szymanski MH, McCown N, Dole KC, Schmechel DE, Strittmatter WJ, Pericak-Vance MA, Roses AD (1993b) Apolipoprotein E epsilon 4 allele distributions in late-onset Alzheimer's disease and in other amyloid-forming diseases Lancet 243 710–711

Schmechel DE, Saunders AM, Strittmatter WJ, Crain BJ, Hulette CM, Joo SH, Pericak-Vance MA, Goldgaber D, Roses AD (1993) Increased amyloid β-peptide deposition in cerebral cortex as consequence of apolipoprotein E genotype in late-onset Alzheimer's disease Proc Natl Acad Sci USA 90 9649–9653

Seeger G, Brauer K, Hartig W, Bruckner G (1994) Mapping of perineuronal nets in the rat brain stained by colloidal iron hydroxide histochemistry and lectin cytochemistry Neuroscience 58 371–388

Skene JHP, Shooter EM (1983) Denervated sheath cells secrete a new protein after nerve injury Proc Natl Acad Sci USA 80 4169–4173

Snipes GJ, McGuire CB, Norden JJ, Freeman JA (1986) Nerve injury stimulates the secretion of apolipoprotein E by nonneuronal cells Proc Natl Acad Sci USA 83 1130–1134

Strittmatter WJ, Saunders AM, Schmechel D, Pericak-Vancer M, Enghild J, Salvesen GS, Roses AD (1993a) Apolipoprotein E High-avidity binding to β-amyloid and increased frequency of type 4 allele in late-onset familial Alzheimer disease Proc Natl Acad Sci USA 90 1977–1981

Strittmatter WJ, Weisgraber KH, Huang DY, Dong L-M, Slavesen GS, Pericak-Vance MA, Schmechel D, Saunders AM, Goldgaber D, Roses AM (1993b) Binding of human apolipoprotein E to synthetic amyloid b peptide isoform-specific effects and implications for late-onset Alzheimer disease Proc Natl Acad Sci USA 90: 8098–8102

Strittmatter WJ, Saunders AM, Goedert M, Weisgraber KH, Dong L-M, Jakes R, Huang DY, Pericak-Vance MA, Schmechel D, Roses AD (1994a) Isoform-specific interactions of apolipoprotein E with tau and phosphorylated tau: implications for Alzheimer Disease. Proc Natl Acad Sci USA 91 11183–11186

Strittmatter WJ, Weisgraber KH, Goedert M, Saunders AM, Huang DY, Corder EH, Dong L-M, Jakes R, Alberts MJ, Gilbert JR, Dole K, Han S-H, Hulette CH, Einstein G, Schmechel DE, Pericak-Vance MA, Roses AD (1994b) Hypothesis microtubule instability and paired helical filament formation in the Alzheimer disease brain as a function of apolipoprotein E genotype Exp Neurol 125 163–171

Tooyama I, Kawamata T, Akiyama H, Moestrupp SK, Gliemann J, McGeer PL (1993) Immunohistochemical study of α_2-macroglobulin receptor in Alzheimer and control postmortem human brain Mol Chem Neuropath 18 153–159

Wisniewski T, Frangione B (1991) Apolipoprotein E a pathological chaperone protein in patients with cerebral and systemic amyloid Neurosci Lett 135 235–238

Apolipoprotein E: Structure, Function, and Possible Roles in Modulating Neurite Extension and Cytoskeletal Activity

*R.W. Mahley***, *B.P. Nathan*, *S. Bellosta*, and *R.E. Pitas*

Summary

Apolipoprotein E (apoE) is a 34000-Da protein that functions in the redistribution of lipids among various tissues. Several observations over the years have suggested that apoE has a role in neurobiology. Most recently it has been associated with the pathogenesis of Alzheimer's disease. One possible mechanism of apoE involvement in this disease may be modulation of the cytoskeleton in neurons and alteration of neurite extension and branching.

Our studies showed that isoforms of apoE exert differential effects on neuronal growth (i.e., neurite extension and branching) in vitro. When combined with a source of lipid, apoE3 stimulated neurite extension in peripheral nervous system neurons (dorsal root ganglia), whereas apoE4 inhibited it. Similar results were obtained with central nervous system neurons (murine neuroblastoma Neuro-2a cells). On the other hand, addition of free apoE3 or apoE4 without lipid to the dorsal root ganglion or Neuro-2a cells had no effect on neurite outgrowth. The apoE4-mediated effects on inhibition of neurite extension require lipoprotein receptor binding and internalization to elicit the response, and the receptor responsible for the apoE4 effect has been identified as the low density lipoprotein (LDL) receptor-related protein.

Recently, studies have been undertaken to determine the effect of apoE on the microtubules of neurons. When neurons were incubated with apoE3 plus a source of lipid, the microtubules were well-formed and extended throughout the cells. On the other hand, neurons incubated with apoE4 plus a source of lipid displayed widely dispersed tubulin and very few well-formed microtubules. Quantitation of monomeric and polymeric tubulin revealed that the tubulin in the apoE4-treated cells was primarily monomeric.

ApoE may impact the stability of microtubules in an isoform-specific manner and affect synaptic remodeling, possibly accounting for the association of the apoE4 isoform with Alzheimer's disease.

* Gladstone Institute of Cardiovascular Disease, Departments of Pathology and Medicine, University of California, San Francisco, CA 94141-9100, USA

A.D. Roses et al. (Eds.)
Apolipoprotein E and Alzheimer's Disease

Introduction

Apolipoprotein E (apoE), a 34000 molecular weight protein that is the product of a single gene on chromosome 19 (for review, see Mahley 1988), exists in three major forms. Its heterogeneity results from amino acid substitutions at residues 112 and 158. The common isoform, apoE3, has cysteine at residue 112 and arginine at residue 158. The apoE4 isoform, which has been linked to Alzheimer's disease, differs from apoE3 only at residue 112, where it possesses arginine. The apoE2 isoform, associated with type III hyperlipoproteinemia (Mahley 1988), differs from apoE3 only at residue 158, where it possesses cysteine. Apolipoproteins E3 and E4 bind normally to the low density lipoprotein (LDL) receptor, whereas apoE2 does not.

By redistributing lipids among the cells of different organs, apoE plays a critical role in lipid metabolism (for review, see Mahley 1988). Although apoE exerts this global transport mechanism in chylomicron and very low density lipoprotein (VLDL) metabolism, it also functions in the local transport of lipids among cells within a tissue. Cells with excess cholesterol and other lipids may release these substances to apoE-lipid complexes or to high density lipoproteins (HDL) containing apoE, which can carry the lipids to cells requiring them for proliferation or repair. The apoE on these lipoprotein particles mediates their interaction and uptake via the LDL receptor or the LDL receptor-related protein (LRP).

A series of observations made over several years suggests a neurobiological role for apoE. First, apoE mRNA is abundant in the brain, where it is synthesized and secreted primarily by astrocytes (Elshourbagy et al. 1985; Boyles et al. 1985; Pitas et al. 1987a). (The brain is second only to the liver in the level of apoE mRNA expression.) Second, apoE-containing lipoproteins are found in the cerebrospinal fluid and appear to play a major role in lipid transport in the central nervous system (Pitas et al. 1987b). In fact, the major cerebrospinal fluid lipoprotein is an apoE-containing HDL. Third, apoE plus a source of cholesterol promotes marked neurite extension in dorsal root ganglion cells in culture (Handelmann et al. 1992). Fourth, apoE levels dramatically increase (about 250-fold) after peripheral nerve injury (Muller et al. 1985; Ignatius et al. 1986). ApoE appears to participate both in the scavenging of lipids generated after axon degeneration and in the redistribution of these lipids to sprouting neurites for axon regeneration and later to Schwann cells for remyelination of the new axons (Boyles et al. 1989; Ignatius et al. 1987).

Most recently, apoE has been implicated in Alzheimer's disease (Saunders et al. 1993; Corder et al. 1993). ApoE is associated with the two characteristic neuropathologic lesions of Alzheimer's disease – extracellular neuritic plaques representing deposits of amyloid beta (Aβ) peptide and intracellular neurofibrillary tangles representing filaments of hyperphosphorylated tau, a microtubule-associated protein (for a review, see McKhann et al. 1984; Selkoe 1991; Crowther 1993; Roses 1994; Weisgraber et al. 1994a,b). Three categories of Alzheimer's disease are early-onset familial disease (occurring before 60 years of age and linked to genes on chromosomes 21 and 14), late-onset familial disease, and sporadic late-onset disease. Both types of late-onset disease have recently been linked to chromosome 19 at the apoE locus. Other results suggest that apoE4 is directly linked to the severity of the disease in late-onset families (Roses 1994).

The neurofibrillary tangles, which are paired helical filaments of hyperphosphorylated tau (for review, see Crowther 1993; Weisgraber et al. 1994a,b), accumulate in the cytoplasm of neurons. Tau is a microtubule-associated phosphoprotein that normally participates in microtubule assembly and stabilization; however, hyperphosphorylation impairs its ability to interact with microtubules.

It has been shown that tau interacts with apoE3 in vitro, but not with apoE4 (Strittmatter et al. 1994). The interaction of apoE3 with tau may prevent tau's hyperphosphorylation, thus allowing it to function normally in stabilizing microtubular structure and function (Weisgraber et al. 1994a; Strittmatter et al. 1994). In the presence of apoE4, tau could become hyperphosphorylated and thus inactive, which could promote the formation of neurofibrillary tangles.

Impact of ApoE on the Cytoskeleton and Neurite Extension in Neurons

In neurons the cytoskeleton functions in neurite extension and retraction. Therefore, our studies have focused on the isoform-specific effects of apoE3 and apoE4 on neurite extension and branching. Our postulate is that apoE modulates the intracellular cytoskeletal apparatus and alters neurite extension and branching. Understanding how the various isoforms of apoE alter the cytoskeleton may shed light on the process of neurofibrillary tangle formation and suggest how apoE may modulate the remodeling of synaptic connections later in life.

Incubation of dorsal root ganglion neurons in culture with β-VLDL or cholesterol alters the neurite growth of these cells compared to that of cells grown in media alone (Handelmann et al. 1992). In the presence of a source of lipid (β-VLDL or free cholesterol), neurite outgrowth is greatly enhanced, specifically due to extensive branching (with little or no increased neurite extension). When rabbit apoE (equivalent to human apoE3 with respect to the occurrence of cysteine at residue 112) was added to this lipid-containing system, however, very significant neurite extension was seen (Handelmann et al. 1992). A source of cholesterol appears to enhance membrane biosynthesis, whereas the addition of rabbit apoE with a source of lipid results in long neuritic extensions and a trimming back of the branches. Stimulation of neurite extension in vivo could promote nerve regeneration or the formation of synaptic connections during neuronal remodeling.

A comparison of the effects of human apoE3 versus human apoE4 showed pronounced differential isoform-specific effects on neurite outgrowth (Nathan et al. 1994a). ApoE3 plus β-VLDL resulted in an increase in neurite extension, whereas apoE4 plus β-VLDL resulted in a marked decrease in both neurite branching and extension. Dorsal root ganglion neurons incubated with apoE4 plus β-VLDL displayed very short, stunted neurites (Nathan et al. 1994a). This was not a toxic effect of apoE4 since replacement of the apoE4-containing media with fresh apoE4-lacking media restored the ability of the neurons to produce neuritic extensions. Furthermore, the apoE3- and apoE4-specific effects were blocked by addition of an antibody against the receptor-binding domain of apoE or by reductive methylation of critical lysine residues, suggesting that the effect of apoE was receptor mediated.

Murine neuroblastoma cells (Neuro-2a) from the central nervous system were used to compare the effects of apoE on the peripheral nervous system neurons described above with the effect on cortical neurons. Cells of both types respond similarly to apoE. When combined with a source of lipid, apoE3 stimulated neurite extension, whereas apoE4 inhibited it (Fig. 1; Nathan et al. 1994b). Addition of free apoE3 or apoE4 without β-VLDL had no effect on neurite outgrowth. These results further suggest that the effect of apoE on neurons requires the lipoprotein receptor-mediated uptake of apoE or a combination of apoE plus a source of lipid. Free of lipid, apoE does not bind to either the LDL receptor or the LRP.

The previous studies were performed by adding large quantities of apoE along with β-VLDL to the cells in culture. To determine whether lower levels of endogenously produced apoE would have an effect on neurite outgrowth from Neuro-2a cells, the neurons were transfected with human apoE cDNA constructs encoding apoE3 or apoE4 (Bellosta et al. 1995). Clones of the transfected cells secreting equal amounts of apoE3 or apoE4 ($\sim$50–60 ng of apoE/mg of cell protein/48 hours) were selected for comparison. The apoE3- and apoE4-secreting cells grown in serum-free control medium displayed a similar degree of limited neurite extension. However, when a source of lipid (β-VLDL) was added to the medium, the cells had a markedly different pattern of growth. The apoE3-secreting cells showed greater neurite extension than did the apoE4-secreting cells (Bellosta et al. 1995). Thus, even very low levels of endogenously produced apoE along with a source of lipid revealed the differential effects of apoE3 versus apoE4. Lipid emulsions of various compositions can be substituted for the β-VLDL and appear to serve as a source of lipid for the cells or as a vehicle for transporting the apoE into a specific intracellular pathway.

Depolymerization of Microtubules Associated with ApoE4 Inhibition of Neurite Outgrowth

Since microtubules play a crucial role in neurite outgrowth, the effect of the two isoforms of apoE on the microtubular architecture of Neuro-2a cells was

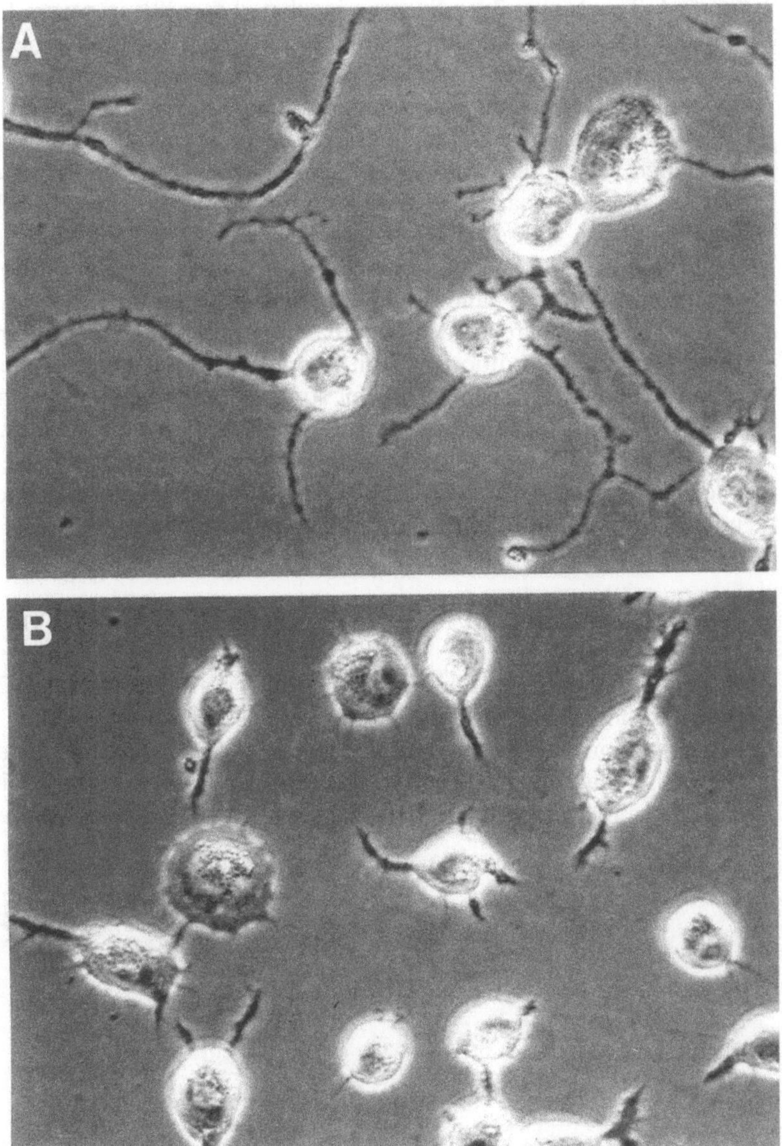

Fig. 1A,B. Isoform-specific effects of apoE on neurite extension in Neuro-2a cells **A** Human apoE3 (30 μg/ml) plus rabbit β-VLDL (very low density lipoproteins, 40 μg cholesterol/ml). **B** Human apoE4 (30 μg/ml) plus rabbit β-VLDL (40 μg cholesterol/ml) The cells were visualized by phase contrast microscopy after 48 h of incubation with the apoE plus β-VLDL *Scale bar*, 20 μm studied. The Neuro-2a cells were incubated for 48h in serum-free medium alone, with β-VLDL alone (40μg cholesterol/ml), or with β-VLDL and either apoE3 (30μg/ml) or apoE4 (30μg/ml). Following incubation, the microtubular architecture of the cells was examined using biochemical, immunocytochemical, and ultrastructural techniques. Immunoblot analysis of tubulin from cells incubated with apoE4 and β-VLDL revealed a significant increase in soluble tubulin and a decrease in polymerized tubulin as compared to cells grown in medium alone, with β-VLDL alone, or with β-VLDL and apoE3. Consistent with the biochemical data, immunocytochemical detection of tubulin in cells incubated with apoE4 and β-VLDL revealed diffuse tubulin staining with few well-organized microtubules, whereas in cells maintained under the other three treatment conditions, a well-formed network of microtubules was observed (Fig. 2). Electron microscopic studies of cells incubated with apoE4 and β-VLDL confirmed the presence of few well-formed microtubules. These data suggest that apoE4 specifically depolymerizes microtubules in Neuro-2a cells and demonstrate that microtubule depolymerization is associated with the inhibition of neurite outgrowth (Nathan et al. 1995a).

ApoE4 Inhibition of Neurite Outgrowth Mediated via the LRP

The inhibitory effect of apoE4 on neurite outgrowth in Neuro-2a cells was observed only in the presence of lipoproteins and was blocked by the binding of a monoclonal antibody to the lipoprotein receptor-binding domain of apoE, suggesting that the interaction of apoE with lipoprotein receptors is important in eliciting the effect. In the present study, we treated cells with reagents that

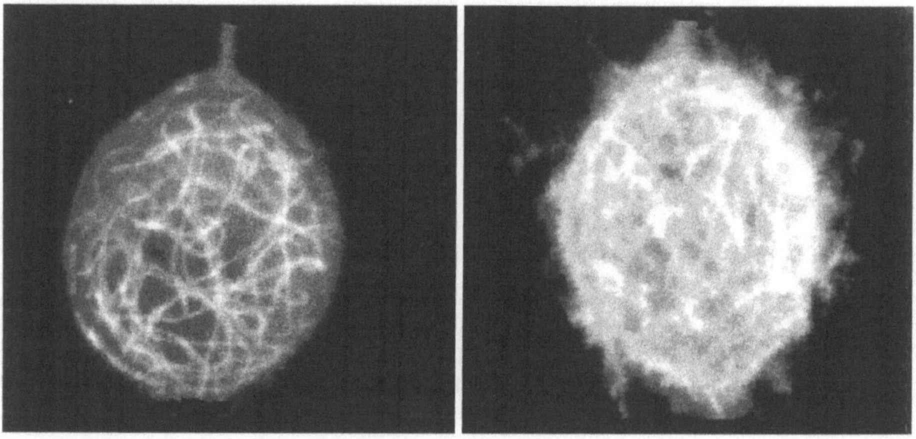

Fig. 2. Effect of apoE on tubulin in neuroblastoma cells Cells were incubated with β-VLDL (very low density lipoproteins) plus apoE3 (*left*) or apoE4 (*right*) The microtubules were observed by immunolocalization using an antibody to β-tubulin and visualized by confocal microscopy

inhibit binding and uptake of lipoproteins by the LRP, but not by the LDL receptor, and examined the effect on apoE4-mediated inhibition of neurite outgrowth. Neuro-2a cells were preincubated for 1 h in serum-free medium alone or in medium containing either heparinase (10 units/ml) or the receptor-associated protein (RAP; 5 μg/ml). Heparinase depletes cell-surface heparan sulfate proteoglycans that bind apoE-enriched lipoproteins prior to internalization via the LRP, whereas the RAP competitively inhibits the binding of lipoproteins to the LRP. Following preincubation, β-VLDL (40 μg cholesterol/ml) and apoE4 (30 μg/ml) were added to the medium, and the cells were incubated for 24 h in the presence of the inhibitors. After this incubation, inhibitors were again added to the medium, and the cells were incubated for an additional 24 h. Heparinase or RAP alone, in the absence of β-VLDL and apoE4, did not have any effect on neurite outgrowth. Consistent with our previous studies, apoE4 in the presence of β-VLDL alone inhibited neurite outgrowth. However, when cells were treated with heparinase or with RAP, the inhibitory effect of apoE4 with β-VLDL on neurite outgrowth was abolished. The data suggest that binding of apoE4-enriched lipoproteins by the LRP is necessary for the inhibitory effect on neurite outgrowth (Nathan et al. 1995b).

Intracellular Accumulation of ApoE3 versus ApoE4

Neuro-2a cells were incubated with β-VLDL and either apoE3 or apoE4 for 48 h, and intracellular apoE was localized by immunofluorescence using confocal microscopy (Nathan et al. 1994b). By optical sectioning through the neuronal cell body and neurites, apoE was visualized on the cell surface and throughout the cell. ApoE3 accumulated in high concentration in the cell bodies around the nucleus and diffusely throughout the neurites within the intracellular compartment (Fig. 3). Although apoE3 clearly is present intracellularly, it is not yet possible to ascertain whether it occurs within the cytoplasm, in association with the cytoskeletal elements, or within vesicles. On the other hand, apoE4 accumulated in the cell bodies and neurites to a much lesser degree; it was localized near the cell surface and to a much lesser extent within the short neurites (Fig. 3). Although we have demonstrated that apoE3 and apoE4 in the presence of β-VLDL are taken up by the cells to a similar extent, apoE3 appears to be retained, whereas apoE4 is not.

Conclusions

In dorsal root ganglion or neuroblastoma cells, apoE3 plus a source of lipid supports and facilitates neurite extension. ApoE3 appears to accumulate widely in cell bodies and neurites, stabilize the cytoskeleton and support neurite elongation, and modulate directly or indirectly microtubule assembly.

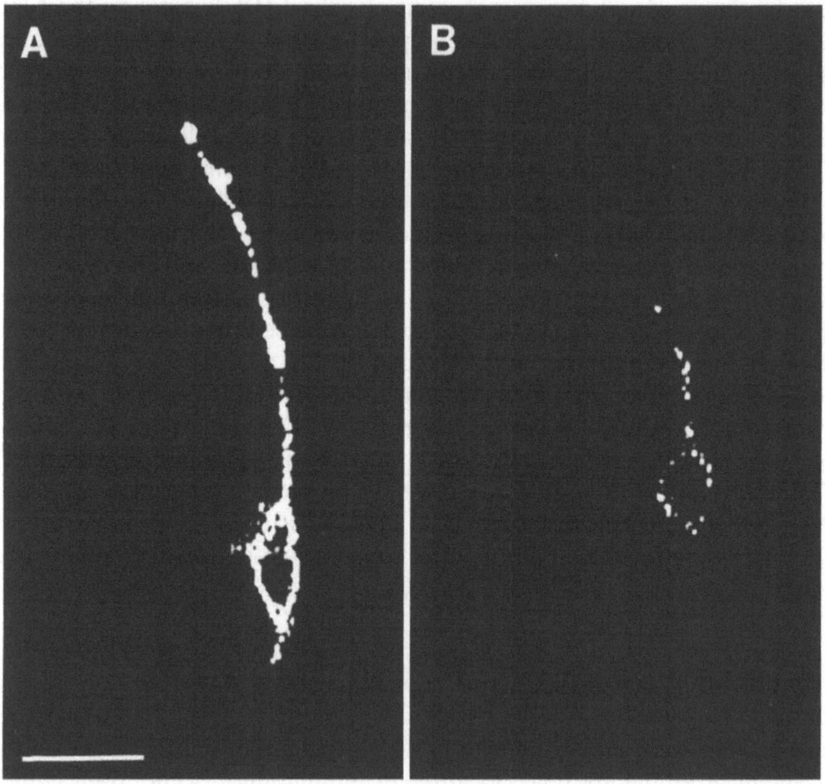

Fig. 3A,B. Intracellular localization of apoE by immunocytochemistry in Neuro-2a cells, as visualized by confocal microscopy **A** Human apoE3 (30 μg/ml) plus rabbit β-VLDL (very low density lipoproteins, 40 μg cholesterol/ml) **B** Human apoE4 (30 μg/ml) plus rabbit β-VLDL (40 μg cholesterol/ml). The Neuro-2a cells were incubated with the apoE plus the β-VLDL for 48 h and then washed and fixed in paraformaldehyde A polyclonal antibody to human apoE was used for immunocytochemistry The antibody–antigen complex was detected by a secondary antibody labeled with Texas red and visualized using a confocal laser scanning microscope *Scale bar*, 40 μm

ApoE4, on the other hand, does not appear to accumulate within neurons or support neurite extension, and may even destabilize the microtubule apparatus. The apoE4 effect appears to be mediated via the LRP pathway. Individuals with apoE4 clearly have normal neuronal development early in life. However, apoE4 may exert its detrimental effects later in life, by not allowing or supporting remodeling of synaptic connections. This effect may be important in the pathogenesis of Alzheimer's disease. Alternatively, apoE4 may contribute to Alzheimer's disease by aiding the formation of dense, complicated, possibly toxic plaques of Aβ peptide. At present, the pathway whereby apoE affects the development of Alzheimer's disease remains speculative.

Acknowledgments. This research was funded in part by NIH Program Project Grant HL41633.

References

Bellosta S, Nathan BP, Orth M, Dong L-M, Mahley RW, Pitas RE (1995) Stable expression and secretion of apolipoproteins E3 and E4 in mouse neuroblastoma cells produces differential effects on neurite outgrowth J Biol Chem 270: 27063–27071

Boyles JK, Pitas RE, Wilson E, Mahley RW, Taylor JM (1985) Apolipoprotein E associated with astrocytic glia of the central nervous system and with nonmyelinating glia of the peripheral nervous system J Clin Invest 76 1501–1513

Boyles JK, Zoellner CD, Anderson LJ, Kosik LM, Pitas RE, Weisgraber KH, Hui DY, Mahley RW, Gebicke-Haerter PJ, Ignatius MJ, Shooter EM (1989) A role for apolipoprotein E, apolipoprotein A-I, and low density lipoprotein receptors in cholesterol transport during regeneration and remyelination of the rat sciatic nerve J Clin Invest 83: 1015–1031

Corder EH, Saunders AM, Strittmatter WJ, Schmechel DE, Gaskell PC, Small GW, Roses AD, Haines JL, Pericak-Vance MA (1993) Gene dose of apolipoprotein E type 4 allele and the risk of Alzheimer's disease in late onset families Science 261 921–923

Crowther RA (1993) Tau protein and paired helical filaments of Alzheimer's disease Curr Opin Struct Biol 3: 202–206

Elshourbagy NA, Liao WS, Mahley RW, Taylor JM (1985) Apolipoprotein E mRNA is abundant in the brain and adrenals, as well as in the liver, and is present in other peripheral tissues of rats and marmosets Proc Natl Acad Sci USA 82. 203–207

Handelmann GE, Boyles JK, Weisgraber KH, Mahley RW, Pitas RE (1992) Effects of apolipoprotein E, β-very low density lipoproteins, and cholesterol on the extension of neurites by rabbit dorsal root ganglion neurons in vitro. J Lipid Res 33. 1677–1688

Ignatius MJ, Gebicke-Harter PJ, Skene JHP, Schilling JW, Weisgraber KH, Mahley RW, Shooter EM (1986) Expression of apolipoprotein E during nerve degeneration and regeneration Proc Natl Acad Sci USA 83 1125–1129

Ignatius MJ, Shooter EM, Pitas RE, Mahley RW (1987) Lipoprotein uptake by neuronal growth cones in vitro Science 236 959–962

Mahley RW (1988) Apolipoprotein E cholesterol transport protein with expanding role in cell biology Science 240: 622–630

McKhann G, Drachman D, Folstein M, Katzman R, Price D, Stadlan EM (1984) Clinical diagnosis of Alzheimer's disease: report of the NINCDS-ADRDA work group under the auspices of department of health and human services task force on Alzheimer's disease Neurology 34 939–944

Muller HW, Gebicke-Harter PJ, Hangen DH, Shooter EM (1985) A specific 37 000-dalton protein that accumulates in regenerating but not in nonregenerating mammalian nerves Science 228 499–501

Nathan BP, Bellosta S, Sanan DA, Weisgraber KH, Mahley RW, Pitas RE (1994a) Differential effects of apolipoproteins E3 and E4 on neuronal growth in vitro Science 264 850–852

Nathan BP, Bellosta S, Mahley RW, Pitas RE (1994b) Apolipoprotein E3- and E4-induced differences in neurite outgrowth are associated with differences in the subcellular localization of apolipoprotein E. Society for Neuroscience 20 (Part 2) 1033 (Abstr.)

Nathan BP, Chang K-C, Bellosta S, Brisch E, Ge N, Mahley RW, Pitas RE (1995a) The inhibitory effect of apolipoprotein E4 on neurite outgrowth is associated with microtubule depolymerization J Biol Chem 270: 19791–19799

Nathan BP, Chang K-C, Bellosta S, Mahley RW, Pitas RE (1995b) The inhibition of neurite outgrowth by apolipoprotein E4 is mediated through the low density lipoprotein receptor-related protein. Society for Neuroscience 21 (Part 2): 1009 (abstract)

Pitas RE, Boyles JK, Lee SH, Foss D, Mahley RW (1987a) Astrocytes synthesize apolipoprotein E and metabolize apolipoprotein E-containing lipoproteins Biochim Biophys Acta 917 148–161

Pitas RE, Boyles JK, Lee SH, Hui DY, Weisgraber KH (1987b) Lipoproteins and their receptors in the central nervous system characterization of the lipoproteins in cerebrospinal fluid and identification of apolipoprotein B,E(LDL) receptors in the brain J Biol Chem 262 14352–14360

Roses AD (1994) The Alzheimer diseases Curr Neurol 14 111–141

Saunders AM, Strittmatter WJ, Schmechel D, St George-Hyslop PH, Pericak-Vance MA, Joo SH, Rosi BL, Gusella JF, Crapper-MacLachlan DR, Alberts MJ, Hulette C, Crain B, Goldgaber D, Roses AD (1993) Association of apolipoprotein E allele ε4 with late-onset familial and sporadic Alzheimer's disease Neurology 43 1467–1472

Selkoe DJ (1991) The molecular pathology of Alzheimer's disease Neuron 6 487–498

Strittmatter WJ, Weisgraber KH, Goedert M, Saunders AM, Huang D, Corder EH, Dong L-M, Jakes R, Alberts MJ, Gilbert JR, Han S-H, Hulette C, Einstein G, Schmechel DE, Pericak-Vance MA, Roses AD (1994) Hypothesis microtubule instability and paired helical filament formation in the Alzheimer disease brain are related to apolipoprotein E genotype Exp Neurol 125 163–171

Weisgraber KH, Roses AD, Strittmatter WJ (1994a) The role of apolipoprotein E in the nervous system Curr Opin Lipidol 5 110–116

Weisgraber KH, Pitas RE, Mahley RW (1994b) Lipoproteins, neurobiology, and Alzheimer's disease structure and function of apolipoprotein E Curr Opin Struct Biol 4 507–515

Apolipoprotein E Role in Maintaining the Integrity of the Aging Central Nervous System

*E. Masliah**, *M. Mallory*, *M. Alford*, *I. Veinbergs*, and *A.D. Roses*

Abstract

Apolipoprotein E (apoE) is involved in the development and regeneration of the central nervous system (CNS). Recent studies have shown that the expression of allele e4 of APOE is a major risk factor for Alzheimer's disease (AD), a neurodegenerative disorder associated with synaptic loss, amyloid deposition and disruption of the neuronal cytoskeleton. In order to better understand the mechanisms by which apoE is involved in AD, it is necessary to understand the physiological role of this molecule in the intact CNS. For this purpose, we studied the fate of synaptic and dendritic populations during aging and after injury to the CNS in apoE-deficient (knockout) mice. In apoE-deficient homozygous mice, there was an age-dependent 15%–40% loss of synaptophysin-immunoreactive nerve terminals and microtubule-associated protein 2-immunoreactive dendrites in the neocortex and hippocampus, when compared to controls. Dendritic alterations were observed as early as 4 months of age. Ultrastructural analysis revealed extensive dendritic vacuolization and disruption of the endomembrane system and cytoskeleton in apoE-deficient homozygous mice. The dendritic damage was accompanied by a decreased immunoreactivity for cytoskeletal markers α and β tubulin (but not kinesin) in the cell bodies and processes of neurons of apoE-deficient mice. In addition, apoE-deficient homozygous mice showed a delay in the patterns of reinnervation of the outer molecular layer of the hippocampal dentate gyrus after perforant pathway transection, and in the neocortex after intraperitoneal kainic acid injection. These results support the contention that apoE might play an important role in maintaining the stability of the synapto-dendritic apparatus, and that altered or deficient functioning of this molecule could underlie the synaptic and cytoskeletal alterations observed in AD.

* University of California, San Diego, School of Medicine, Department of Neurosciences, La Jolla, CA 92093-0624, USA

A D Roses et al (Eds)
Apolipoprotein E and Alzheimer's Disease

Introduction

During the normal aging process, the CNS is exposed to a multitude of endogenous challenges including oxygen radicals and excitotoxins. Neurons within the neocortex and limbic system are particularly vulnerable to these factors and appear to degenerate earlier compared to other neuronal groups (Hof and Morrison 1994). However, within these regions neuroprotection is possible if specific neuronal groups express agents that: (a) reduce or buffer calcium influx (i.e., calcium-binding proteins; Mattson et al. 1993a,b), (b) scavenge oxygen radicals (i.e., superoxide dismutase), and (c) promote compensation by sprouting of unaffected collaterals (i.e., neurotrophic factors; Mori 1993; Mucke et al. 1995). Among them, recent studies have shown that apoE is involved in remodeling and regeneration of the nervous system (Boyles et al. 1990; Handelmann et al. 1992; Masliah et al. 1995; Nathan et al. 1994; Pitas et al. 1987; Poirier et al. 1993; Poirier 1994), suggesting that it could play an important role in maintaining the integrity of the CNS during aging.

In the CNS, apoE is primarily synthesized and secreted by astrocytes (Boyles et al. 1985; Pitas et al. 1987; Poirier et al. 1991) and macrophages (Pitas et al. 1981), and its synthesis is increased after injury (Poirier 1994). Non-esterified cholesterol released after injury is esterified and transported by apoE to neurons undergoing reinnervation and taken up through the low density lipoprotein-receptor (LDL-R) pathway, where it is used as a precursor for the synthesis of new synaptic terminals (Poirier et al. 1993). Furthermore, the LDL-R is present at high concentrations in regenerating peripheral nerve processes (Boyles et al. 1990), suggesting that apoE mediates the delivery of lipids to regenerating axons (Handelmann et al. 1992). Further supporting the possibility that apoE might play an important role in the nervous system, recent studies have shown that age-associated neurodegenerative disorders could be associated with the APOE genotype. For example, in AD, a disorder associated with synaptic loss, amyloid deposition and disruption of the neuronal cytoskeleton (for review, see Terry et al. 1994), the expression of the ε4 allele of APOE is a major risk factor (Saunders et al. 1993).

Taken together these studies suggest that apoE plays an important physiological role within the CNS; however, the precise mechanisms are not yet known. To better understand the mechanisms by which apoE helps to maintain the stability of the nervous system, we studied the integrity of the synapto-dendritic apparatus in aging apoE-deficient (knockout) mice (E. Masliah et al., submitted for publication), as well as in APOE knockout mice subjected to entorhinal cortex lesions (Masliah et al. 1994, 1995) or kainic acid challenge. This study reviews in vivo evidence that supports the contention at apoE might play an important role in maintaining the integrity of the CNS, during aging and after injury, by preserving the stability of the neuronal cytoskeleton.

Neurodegeneration in the CNS of APOE Knockout Mice During Aging

Young adult, adult and aged apoE-deficient homozygous mice (C57BL/6J) showed significant synaptic and dendritic alterations in the neocortex and limbic system when compared to age- and line-matched apoE-deficient heterozygous and wild-type control mice (Fig. 1A,B; E. Masliah et al., submitted). ApoE-deficient homozygous mice showed an age-dependent 15%–40% decrease in synaptophysin (SYN, marker of presynaptic terminals) and microtubule-associated protein 2 (MAP2, dendritic marker) immunoreactivity in the hippocampus and neocortex when compared to age-matched controls ($p <$ 0.01, one way ANOVA, n = 3 mice per group and age; Fig. 1). Analysis of double-immunolabeled sections with the laser scanning confocal microscope (LSCM) also showed that, in apoE-deficient homozygous mice, the dendrites displayed a widespread vacuolization and they were tortuous, dilated and irregular (Fig. 2). These dendritic alterations were not observed in age-matched apoE-deficient heterozygous and wild-type mice (Fig. 2a,b). Both axonal and dendritic processes occasionally appeared to be thinner when

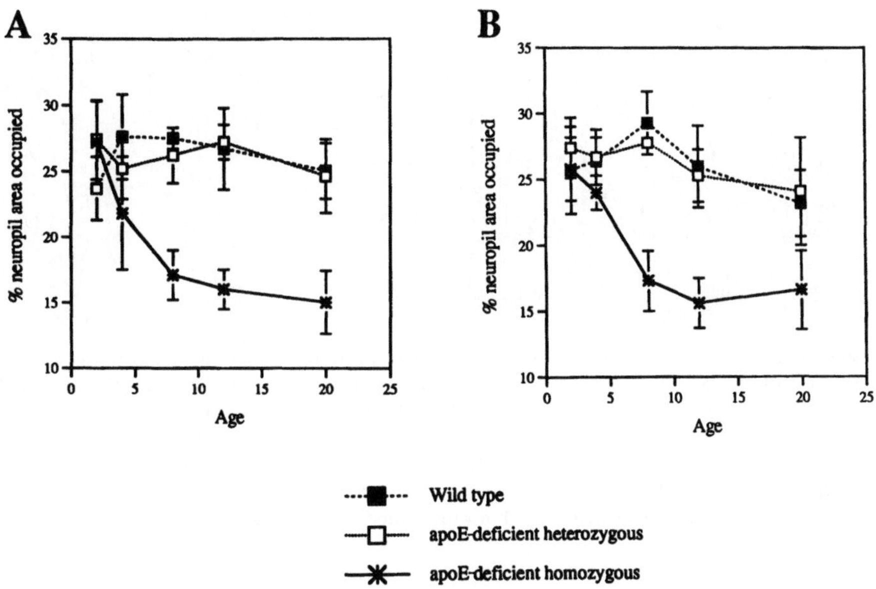

Fig. 1A,B. Computer-aided quantitative analysis of the percentage area of the neuropil occupied by microtubule-associated protein 2 (MAP2)-immunoreactive dendrites (**A**) and synaptophysin (SYN)-immunoreactive presynaptic terminals (**B**). ApoE-deficient homozygous mice displayed a significant, age-dependent 15%–40% loss of MAP2- and SYN-immunoreactive elements in the frontal cortex One way analysis of variance (ANOVA) showed significant differences between apoE-deficient homozygous mice and age-matched controls at the $p < 0.01$ level in the neocortex and at $p < 0 05$ (n = 3 mice per age group and genotype)

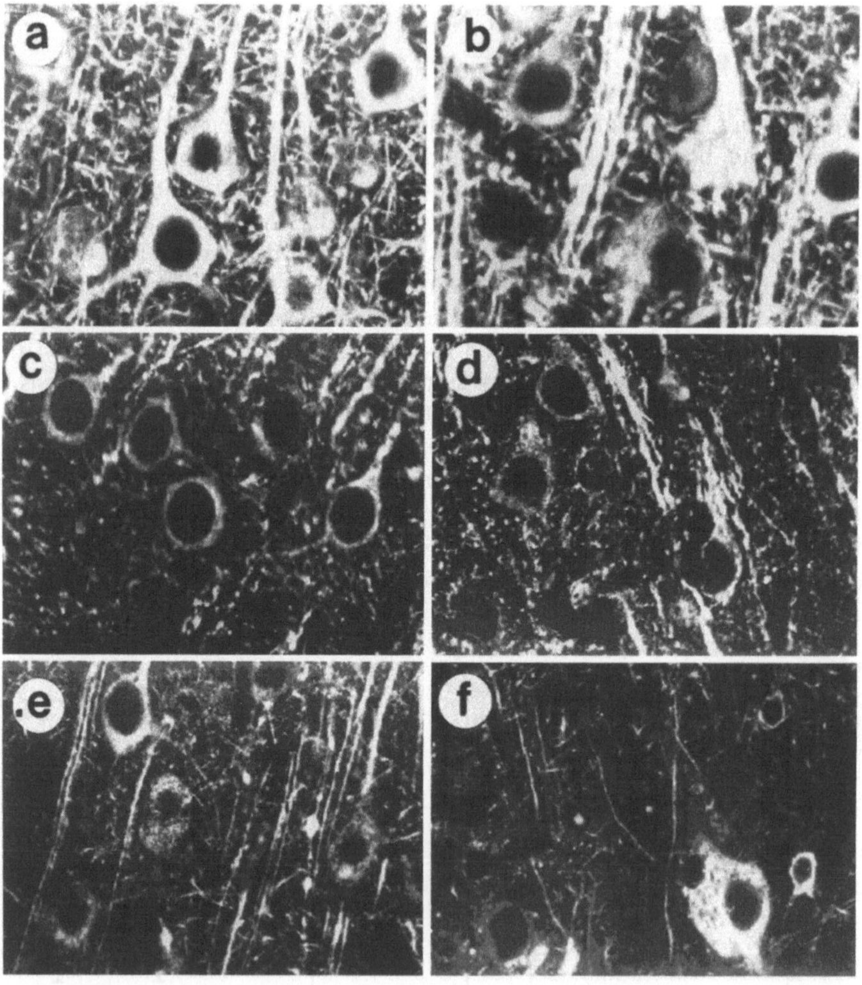

Fig. 2a–f. Dendritic alterations in the neocortex of aged apoE-deficient homozygous mice assessed with the laser scanning confocal microscope (LSCM) in sections labeled with an antibody against microtubule-associated protein 2 (MAP2) Compared to wild-type (**a**) and apoE-deficient heterozygous (**b**) control mice, apoE-deficient homozygous mice from Rockefeller University/ UCSD (**c**) and from UNC/Jackson Labs (**d**) displayed significant dendritic alterations In addition to the dendritic vacuolization and loss of complexity, other aged apoE-deficient homozygous mice showed dendritic processes that were thin with dilated segments (**e,f**)

compared to controls (Fig. 2e,f). The synaptic and dendritic alterations were observed both in apoE-deficient homozygous mice bred at UCSD (from founders kindly provided by Dr. Jan Breslow, Rockefeller University; Plump et al. 1992), as well as in mice independently bred at UCSD by Drs. Steinberg and Palinsky (Fig. 2c–f). Further confirmation of the results was obtained by analyzing apoE-deficient homozygous (C57BL/6J) mice obtained from a commercial source. These mice were originally generated and characterized by Maeda and colleagues (Piedrahita et al. 1992; Zhang et al. 1992) from University of North Carolina at Chapel Hill, and are available from Jackson Laboratories. Analysis with the LSCM has shown similar dendritic alterations in aged (12 months old) apoE-deficient homozygous mice, compared to controls (Fig. 2c-f). In addition, both immunoblot and electron microscopic analysis showed an average 25% decrease in synaptic and dendritic markers in apoE-deficient homozygous mice, when compared to controls (Fig. 3).

Further ultrastructural analysis showed that, in the apoE-deficient homozygous mice, microtubules were disarrayed and fragmented (Fig. 3) and that the endomembrane system was distended and vacuolized and inclusions were detected in spines and secondary segments of the dendrites (Fig. 3). Occasionally some presynaptic terminals appeared dilated and with reduction in the numbers of synaptic vesicles. These findings, in combination with alterations observed in the dendrites with MAP2, suggest that apoE might maintain the integrity of the synapto-dendritic complex by interacting with the cytoskeleton. In this regard, immunocytochemical analysis with antibodies against β-tubulin and tyrosinated and non-tyrosinated α-tubulin showed an overall decrease in immunoreactive neurons, accompanied by fragmentation and disruption of the neuritic processes in aged apoE-deficient homozygous mice (E. Masliah et al., submitted). Currently experiments are underway to assess whether infusion of apoE into the lateral ventricles utilizing osmotic pumps rescues or enhances the cytoskeletal and synapto-dendritic alterations observed in APOE knockout mice.

Role of ApoE in Recovery from CNS Injury

To assess the role of apoE in the patterns of denervation and reinnervation of the CNS, we challenged apoE-deficient homozygous and control mice with either aspiration of the perforant pathway (Masliah et al. 1995) or intraperitoneal kainic acid injection (20 mg/kg). Interruption of perforant pathway by aspiration denervates the projections of the entorhinal cortex into the outer molecular layer (OML) of the hippocampal dentate gyrus (Masliah et al. 1991; Matthews et al. 1976). This paradigm represents a useful model to study the influence of apoE, as well as other trophic factors, on the patterns of denervation and reinnervation of the CNS (Masliah et al. 1991; Poirier et al. 1993). Three days post aspiration of the perforant path, there was a significant 50% decrease in the numbers of SYN-immunoreactive terminals in the OML of control mice, while apoE-deficient homozygous mice showed a 60% loss of SYN-immunoreactive terminals in the OML. Thirty days after lesion, control mice displayed a normal pattern of recovery in the numbers of SYN-immunoreactive nerve terminals. In contrast, apoE-deficient homozygous mice displayed only a mild recovery 30 days post lesion. Analysis of sections

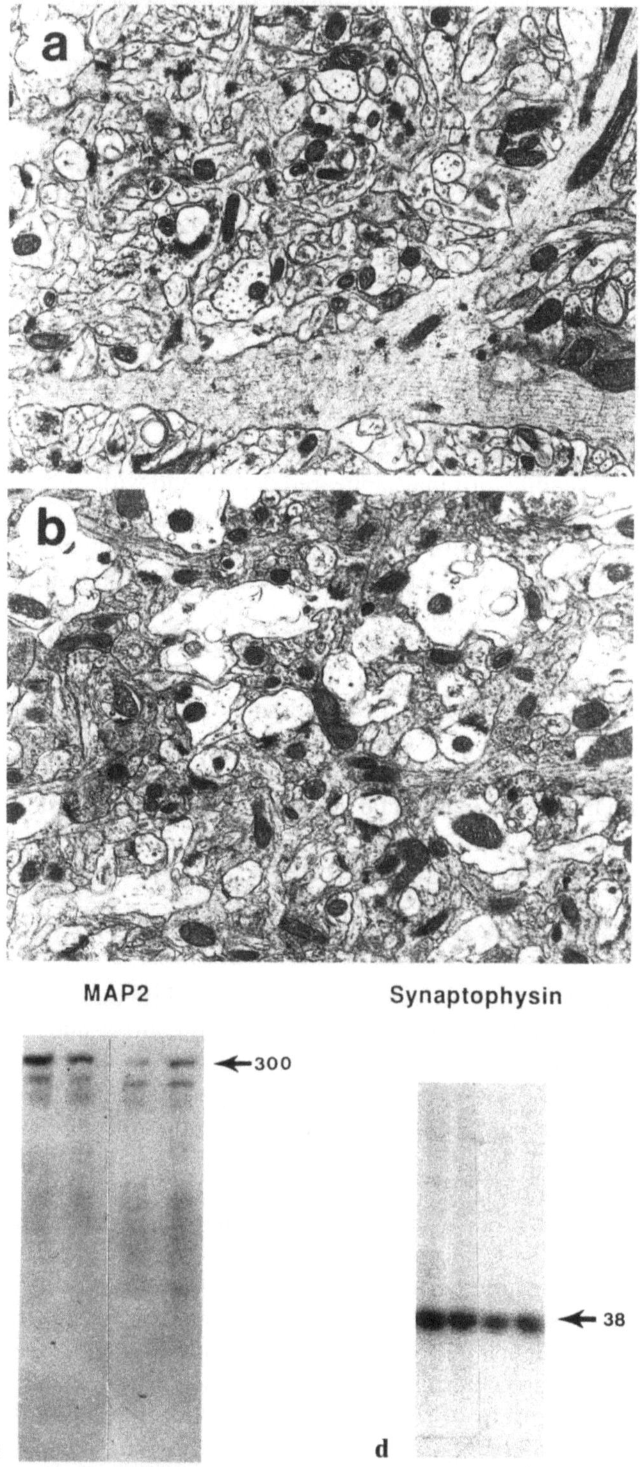
a
b
MAP2
300
c
c c e e
Synaptophysin
38
d
c c e e

immunolabeled with anti-glial fibrillary acidic protein (GFAP, astroglial marker) showed a similar increase in astrocytic immunolabeling at 0, 3, and 30 days post lesion in both control and apoE-deficient homozygous mice (Fig. 4).

Although perforant pathway transection allows the study of regeneration during the course of several weeks, excitatory damage to the CNS by intraperitoneal injection of kainic acid (or similar agents) results in an acute and severe lesion that challenges the ability of the CNS to regenerate (Andersson et al. 1991; Garcia-Ladona and Gombos 1993; Strain and Tasker 1991; Urca and Urca 1990). Five days after kainic acid treatment, apoE-deficient homozygous mice displayed a 65% loss of MAP2-immunoreactive dendritic elements in the neocortex, accompanied by extensive disruption and vacuolization of the neuropil, compared to saline-treated apoE-deficient homozygous mice (Fig. 5). In contrast, kainic acid-treated control mice showed an average 35% loss in MAP2-immunoreactivity compared to saline-treated controls (Fig. 5). The hippocampal pyramidal layer also presented significant alterations after kainic acid; however, the damage observed in the neocortex was more prominent. These observations contrast with studies in the rat, where there was a more circumscribed damage to the hippocampus after kainic acid injection. Analysis of sections immunolabeled with anti-GFAP showed that, compared to saline-treated control mice, saline-treated apoE-deficient homozygous mice presented a moderate astroglial reaction (Fig. 6). In contrast, after kainic acid injection, both groups showed a significant astrogliotic reaction (Fig. 6). Fourteen days after lesion, kainic acid-treated control mice displayed a slight trend toward recovery, whereas apoE-deficient homozygous mice did not show synaptic and dendritic improvement . These results suggest that the extensive synapto-dendritic damage in the kainic acid-treated apoE-deficient homozygous mice represents an additive effect, since originally the intact CNS of these animals display an average 30% loss. Therefore, kainic acid further enhances the damage, bringing it below the threshold of recovery. Furthermore, consistent with previous reports (Andersson et al. 1991; Garcia-Ladona and Gombos 1993; Strain and Tasker 1991; Urca and Urca 1990), recovery after kainic acid lesion in the control mice appears to be very slow and incomplete.

◀

Fig. 3a–d. Ultrastructural and immunochemical analysis of the synaptic content in the frontal cortex of age-matched, 8-month-old mice. Low power view (×5000) electron micrograph of the neuropil in control mouse (**a**) and in apoE-deficient homozygous mouse (**b**), showing an extensive vacuolization in the latter. Quantitative analysis of synaptic contact from the electron micrographs showed a 30% loss in apoE-deficient homozygous mice compared to controls ($p < 0.05$, one-way analysis of variance, ANOVA, $n = 3$ mice per group). Consistent with the electron microscopic results, immunodot-blot analysis with antibodies against microtuble-associated protein 2 (*MAP2*) and synaptophysin showed a decrease in this synaptic marker in apoE-deficient homozygous mice (**d**) compared to the control group (**c**)

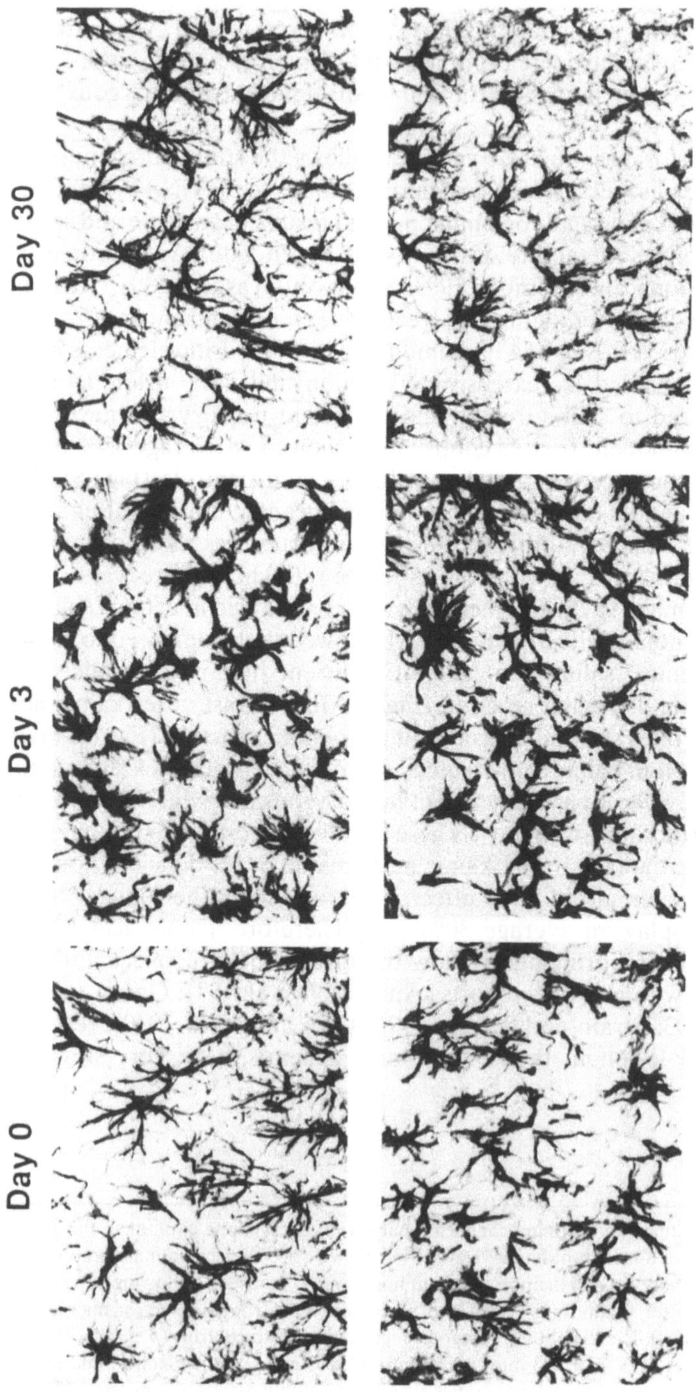

Fig. 4. Astroglıal reactıon ın control and apoE-deficıent mice after entorhınal cortex deafferentatıon. Sectıons were ımmunolabeled wıth an antıbody against glıal fibrıllary acıdıc proteın (GFAP) and analyzed at 0, 3, and 30 days post-lesion

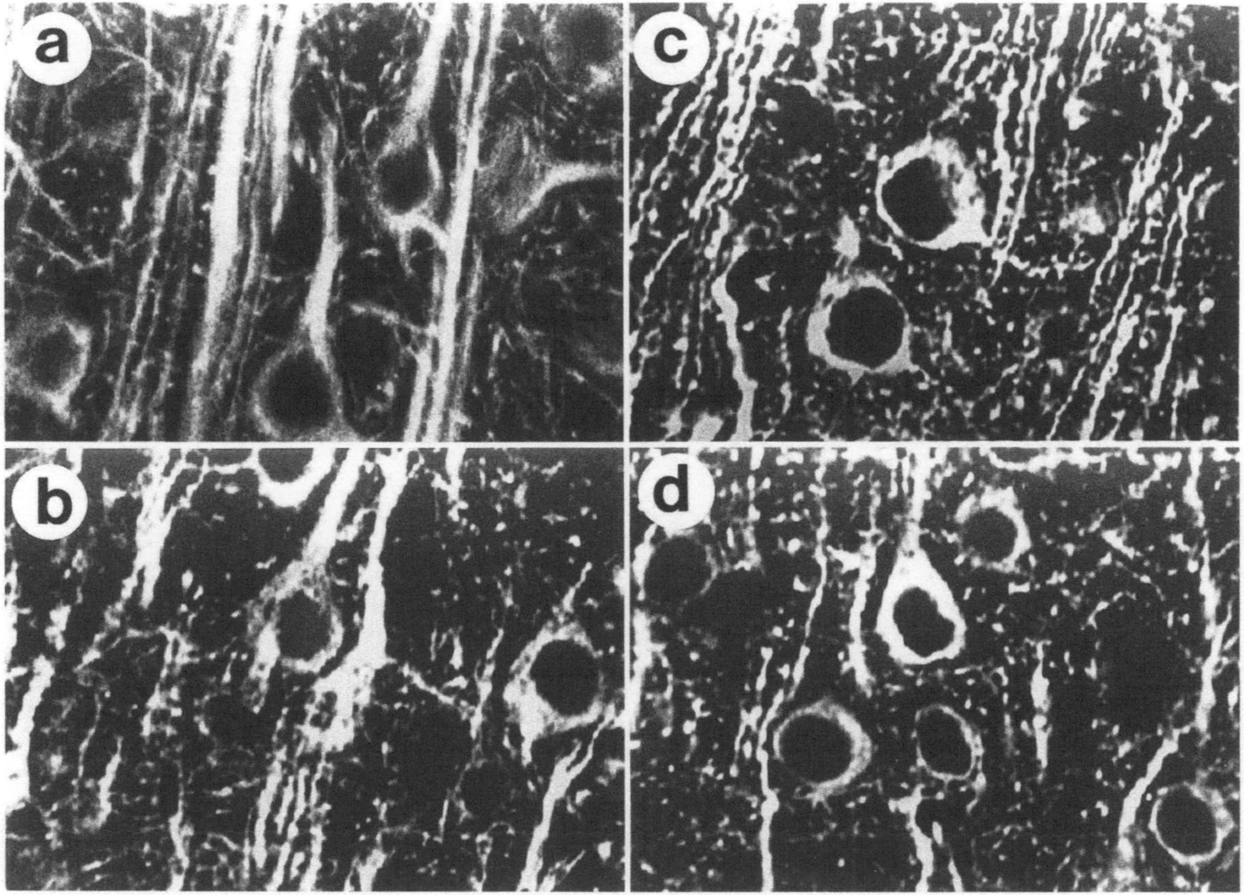

Fig. 5a–d. Dendritic alterations in the neocortex after kainic acid lesion Administration of **a,c** saline solution and **b,d** kainic acid. ApoE-deficient homozygous mice (**c,d**) displayed a greater damage after intraperitoneal kainic acid administration compared to

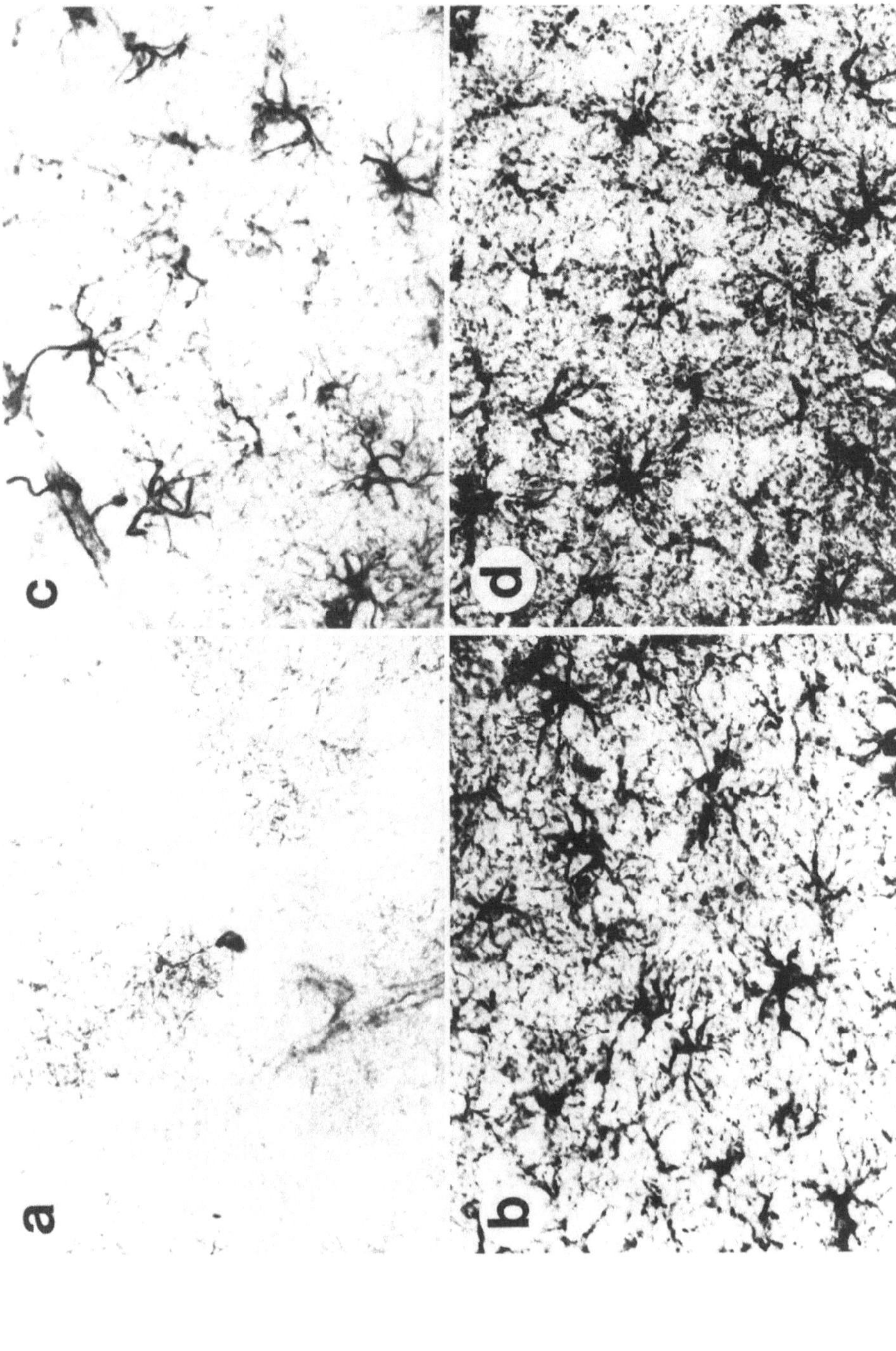

Fig. 6a–d. Astroglotic reaction in the neocortex 5 days after kainic acid lesion. Administration of a, c saline solution and b, d kainic acid

Discussion

These results are consistent with previous studies that have indicated a possible role for apoE in the plasticity of CNS (Boyles et al. 1990; Handelmann et al. 1992; Poirier et al. 1991, 1993; Poirier 1994), suggesting that apoE might be critical in maintaining the integrity of the CNS during aging and after lesions (Masliah et al. 1995; Poirier et al. 1993). In this regard, it has been recently shown that non-esterified cholesterol is released, esterified and transported by apoE to neurons undergoing regeneration after entorhinal cortex lesions, where it is used as a precursor for the formation of new synapses (Poirier et al. 1993). These results contrast with previous studies showing that, although apoE production is increased in mice after injury of the peripheral nervous system (PNS; Boyles et al. 1990), nerve regeneration occurs in the absence of apoE in APOE knockout mice (Popko et al. 1993). Although in the PNS apoE deficiency might be compensated for by overproduction of apoD, A-I, and A-IV (Boyles et al. 1990; Popko et al. 1993), in the CNS apoE appears to be indispensable for the maintenance and regeneration after lesion (Masliah et al. 1995; Poirier et al. 1993). In AD there is extensive neuronal and synapse loss in the limbic system and neocortex, accompanied by amyloid deposition and tangle formation (for a review, see Terry et al. 1994). Taken together these findings, in conjunction with recent studies showing that approximately 64% of AD cases are associated with the presence of allele ε4 of APOE (Saunders et al. 1993), suggest that an abnormally functioning apoE, in addition to contributing to amyloid deposition and tangle formation (Diedrich et al. 1991; Namba et al. 1992; Rebeck et al. 1993; Strittmatter et al. 1993a, 1993b, 1994), might lead to abnormal sprouting and synapse loss in AD.

ApoE-deficient homozygous mice displayed an age-dependent disruption of the synaptic and dendritic organization of the neocortex and limbic system that was accompanied by significant cytoskeletal alterations. These neuritic alterations could be related to either the combined effect of apoE deficit and aging or could be exclusively secondary to the aging process. Previous studies in C57BL/6N mice have shown that, during aging, medium spiny I striatal neurons display retraction of their dendritic tree, which is correlated with poor motor performace (McNeill et al. 1990). Similar alteration has been reported in age-dependent studies in the rat superior cervical ganglion (Andrews et al. 1994) and motor neurons (O'Hanlon and Lowrie 1993). However, none of these previous studies has shown dendritic alterations during aging similar to the ones we observed in the apoE-deficient homozygous mice, indicating a possible role for apoE in maintaining the neuritic structure. The mechanisms by which apoE protects the CNS during aging and after lesion are not yet clear. Recent studies have suggested that apoE might achieve this effect by: (a) stabilizing calcium influx into the neuron (Hartmann et al. 1994), (b) transporting cholesterol to the regenerating plasma membrane (Poirier et al. 1993;

Poirier 1994), and/or (c) stabilizing the neuronal cytoskeleton (Masliah et al. 1995).

The results suggesting a relationship between apoE and the neuronal microtubular cytoskeleton are surprising in view of the fact that apoE is mainly synthesized by astroglial cells (Boyles et al. 1985); thus, the mechanisms by which apoE could be incorporated into neurons to regulate its cytoskeletal function are not yet clear. In this regard, immunocytochemical studies have shown that receptors for apoE are located in neuronal cell bodies and dendrites (Rebeck et al. 1993), suggesting that apoE could be incorporated into the neuron through this mechanism. Furthermore, apoE can be internalized by neurons (Ignatius et al. 1987) and may serve as a major cholesterol carrier to the neurons (Handelmann et al. 1992). Supporting the possibility that apoE plays a role in stabilizing the cytoskeleton via cholesterol uptake, recent studies have shown that this lipid is necessary for the formation of microfilaments in rat astrocytes (Ito et al. 1994), microtubule polymerization in thyroid epithelial cells (Bifulco et al. 1993) and cytoskeletal organization in human carcinoma cells (Ludes et al. 1993). In AD, a disorder characterized by an extensive disruption of the neuronal cytoskeletal system (McKee et al. 1991), apoE immunoreactivity has been shown to be associated with pyramidal neurons and neurofibrillary tangles (Han et al. 1994a,b; Namba et al. 1991, 1992; Strittmatter et al. 1994).

Recent in vitro studies have shown isoform-specific, high-avidity binding between apoE3 and microtubule-associated protein τ (Strittmatter et al. 1994). In contrast apoE4 does not bind τ (Strittmatter et al. 1994), further suggesting that apoE might play a role in delaying paired helical filament formation and sequestering τ for effective microtubular interactions. In addition to τ, MAP2 is also a member of the family of neuronal microtubule-associated proteins that promote microtubule assembly and stabilizatior (Goedert et al. 1991). In a way similar to τ, apoE3 (but not apoE4) binds MAP2c with high avidity, forming a complex that resists dissociation by boiling in sodium dodecyl sulfate. The isoform-specific interactions of apoE with MAP2c and tau appear to play a central role in maintaining the structural stability of the synapto-dendritic cytoskeleton and endomembrane system in the CNS. In AD, it is possible that apoE dysfunction could affect τ phosphorylation of microtubule-associated proteins and microtubule assembly leading to neurodegeneration.

Acknowledgments. The authors would like to thank Drs. Breslow, Palinsky and Steinberg for providing the mice necessary for the present study. This work was supported by grants AG05131, AG10689 and with funding from the Neurosciences Education and Research Foundation, the Alzheimer's Association and the Sam and Rose Stein Institute for Research on Aging (EM); The Joseph and Kathleen Bryan Alzheimer's Disease Research Center (AG05128) and the LEAD Award (AG07922) (ADR).

References

Andersson P-B, Perry VH, Gordon S (1991) The kinetics and morphological characteristics of the macrophage-microglial response to kainic acid-induced neuronal degeneration Neuroscience 42: 210–214

Andrews TJ, Li D, Halliwell J, Cowen T (1994) The effect of age on dendrites in the rat superior cervical ganglion J Anat 184 111–117

Bifulco M, Laezza C, Aloj SM, Garbi C (1993) Mevalonate controls cytoskeleton organization and cell morphology in thyroid epithelial cells J Cell Physiol 155· 340–348

Boyles JK, Pitas RE, Wilson E, Mahley RW, Taylor JM (1985) Apolipoprotein E associated with astrocytic glia of the central nervous system and with nonmyelinating glia of the peripheral nervous system J Clin Invest 76 1501–1513

Boyles JK, Notterpek LM, Anderson LJ (1990) Accumulation of apolipoproteins in the regenerating and remyelinating mammalian peripheral nerve J Biol Chem 265 17805–17815

Diedrich JF, Minnigan H, Carp RI, Whitaker JN, Race R, Frey W II, Haase AT (1991) Neuropathological changes in scrapie and Alzheimer's disease are associated with increased expression of apolipoprotein E and cathepsin D in astrocytes J Virol 65 4759–4768

Garcia-Ladona FJ, Gombos G (1993) Regional distribution of transient [^{3}H] kainic acid-binding sites in the central nervous system of the developing mouse an autoradiographic study Neurosci Lett 151 209–213

Goedert M, Crowther RA, Garner CC (1991) Molecular characterization of microtubule-associated proteins tau and MAP2 Trends Neurosci 14. 193–199

Han S-H, Einstein G, Weisgraber KH, Strittmatter WJ, Saunders AM, Vance-Pericak M, Roses AD, Schmechel DE (1994a) Apolipoprotein E is localized to the cytoplasm of human cortical neurons A light and electron microscopic study J Neuropathol Exp Neurol 53 535–544

Han S-H, Hulette C, Saunders AM, Einstein G, Pericak-Vance M, Strittmatter WJ, Roses AD, Schmechel DE (1994b) Apolipoprotein E is present in hippocampal neurons without neurofibrillary tangles in Alzheimer's disease and in age-matched controls Exp Neurol 128 13–26

Handelmann GE, Boyles JK, Weisgraber KH, Mahley RW, Pitas RE (1992) Effects of apolipoprotein E, β-very low density lipoproteins, and cholesterol on the extension of neurites by rabbit dorsal root ganglion neurons in vitro J Lipid Res 33· 1677–1688

Hartmann H, Eckert A, Muller WE (1994) Apolipoprotein E and cholesterol affect neuronal calcium signalling the possible relationship to beta-amyloid neurotoxicity. Biochem Biophys Res Commun 200 1185–1192

Hof PR, Morrison JH (1994) The cellular basis of cortical disconnection in Alzheimer disease and related dementing conditions In Terry RD, Katzman R, Bick KL (eds) Alzheimer disease Raven Press, New York, pp 197–230

Ignatius MJ, Shooter EM, Pitas RE, Mahley RW (1987) Lipoprotein uptake by neuronal growth cones in vitro Science 236 959–962

Ito J, Masuda M, Tanaka R (1994) Sialosylcholesterol effects on reconstitution of microfilament and glia filament J Neurochem 62 235–239

Ludes B, Schmit AC, Cremel G, Lambert AM, Hubert P, Jacqmin D, Bollack C, Staedel C (1993) Influence of cholesterol derivatives on cytoskeletal organization of human carcinoma cells Eur Urol 23 490–501

Masliah E, Fagan AM, Terry RD, DeTeresa R, Mallory M, Gage FH (1991) Reactive synaptogenesis assessed by synaptophysin immunoreactivity is associated with GAP-43 in the dentate gyrus of the adult rat. Exp Neurol 113· 131–142

Masliah E, Mallory M, Alford M, Mucke L (1994) Abnormal synaptic regeneration in hAPP transgenic and APOE-knockout mice Neurobiol Aging 15· S11

Maslıah E, Mallory M, Alford M, Ge N, Mucke L (1995) Abnormal synaptıc regeneratıon ın hAPP695 transgenıc and APOE knockout mıce In· Iqbal K, Mortımer JA, Wınblad B, Wısnıewskı HM (eds) Research advances ın Alzheımer's dısease and related dısorders John Wıley & Sons Ltd, pp 405–414

Matthews DA, Cotman C, Lynch G (1976) An electron mıcroscopıc study of lesıon-ınduced synaptogenesıs ın the dentate gyrus of the adult rat II Reappearance of morphologıcally normal synaptıc contacts Braın Res 115 23–41

Mattson MP, Cheng B, Culwell AR, Esch FS, Lıeberburg I, Rydel RE (1993a) Evıdence for excıtoprotectıve and ıntraneuronal calcıum-regulatıng roles for secreted forms of the β-amyloıd precursor proteın Neuron 10 243–254

Mattson MP, Cheng B, Smıth-Swıntosky VL (1993b) Mechanısms of neurotrophıc factor protectıon agaınst calcıum- and free radıcal- medıated excıtotoxıc ınjury Implıcatıons for treatıng neurodegeneratıve dıssorders Exp Neurol 124 89–95

McKee AC, Kosık KS, Kowall NW (1991) Neurıtıc pathology and dementıa ın Alzheımer's dısease Ann Neurol 30 156–165

McNeıll TH, Koek LL, Brown SA, Rafols JA (1990) Quantıtatıve analysıs of age-related dendrıtıc changes ın medıum spıny I (MSI) strıatal neurons of C57BL/6N mıce Neurobıol Agıng 11 573–550

Morı N (1993) Toward understandıng of the molecular basıs of loss of neuronal plastıcıty ın ageıng. Age Ageıng 22 S5–S18

Mucke L, Abraham CR, Ruppe MD, Rockensteın EM, Toggas SM, Alford M, Maslıah E (1995) Protectıon agaınst HIV-1 gp 120-ınduced braın damage by neuronal overexpressıon of human amyloıd precursor proteın (hAPP) J Exp Med 181 1551–1556

Namba Y, Tomonaga M, Kawasakı H, Otomo E, Ikeda K (1991) Apolıpoproteın E ımmunoreactıvıty ın cerebral amyloıd deposıts and neurofibrıllary tangles ın Alzheımer's dısease and kuru plaque amyloıd ın Creutzfeld-Jakob dısease Braın Res 541· 163–166

Namba Y, Tsuchıya H, Ikeda K (1992) Apolıpoproteın B ımmunoreactıvıty ın senıle plaque and vascular amyloıds and neurofibrıllary tangles ın the braıns of patıents wıth Alzheımer's dısease Neuroscı Lett 134. 264–266

Nathan BP, Bellosta S, Sanan DA, Weısgraber KH, Mahley RW, Pıtas RE (1994) Dıfferentıal effects of apolıpoproteıns E3 and E4 on neuronal growth ın vıtro Scıence 264: 850–852

O'Hanlon GM, Lowrıe MB (1993) Evıdence for age-dependent changes ın motoneuron dendrıtıc morphology followıng neonatal nerve-crush ın the rat Neuroscıence 57 995–1005

Pıedrahıta JA, Zhang SH, Hagaman JR, Olıver PM, Maeda N (1992) Generatıon of mıce carryıng a mutant apolıpoproteın gene ınactıvated by gene targetıng ın embryonıc stem cells. Proc Natl Acad Scı USA 89 4471–4475

Pıtas RE, Innerarıty TL, Weınsteın JN, Mahley RW (1981) Acetoacetylated lıpoproteıns used to dıstınguısh fibroblasts from macrophages ın vıtro by fluorescence mıcroscopy Arterıosclerosıs 1 177–185

Pıtas RE, Boyles JK, Lee SH, Foss D, Mahley RW (1987) Astrocytes synthesıze apolıpoproteın E and metabolıze apolıpoproteın E-contaınıng lıpoproteıns. Bıochem Bıophys Res Commun 917: 148–161

Plump AS, Smıth JD, Hayek T, Aaalto-Setala K, Walsh A, Verstuyft JG, Rubın EM, Breslow JL (1992) Severe hypercholesterolemıa and atherosclerosıs ın apolıpoproteın E-deficıent mıce created by homologous recombınatıon ın ES cells Cell 71 343–353

Poırıer J (1994) Apolıpoproteın E ın anımal models of CNS ınjury and ın Alzheımer's dısease Trends Neuroscı 17 525

Poırıer J, Hess M, May PC, Fınch CE (1991) Apolıpoproteın E-and GFAP-RNA ın hıppocampus durıng reactıve synaptogenesıs and termınal prolıferatıon Molec Braın Res 11: 97–106

Poırıer J, Baccıchet A, Dea D, Gauthıer S (1993) Cholesterol synthesıs and lıpoproteın reuptake durıng synaptıc remodelıng ın hıppocampus ın adult rats Neuroscı 55 81–90

Popko B, Goodrum JF, Bouldın TW, Zhang SH, Maeda N (1993) Nerve regeneratıon occurs ın the absence of apolıpoproteın E ın mıce J Neurochem 60 1155–1158

Rebeck GW, Reiter JS, Strickland DK, Hyman BT (1993) Apolipoprotein E in sporadic Alzheimer's disease: allelic variation and receptor interactions. Neuron 11 575–580

Saunders AM, Strittmatter WJ, Schmechel D, St George-Hyslop PH, Pericak-Vance MA, Joo SH, Rosi BL, Gusella JF, Crapper-MacLachlan DR, Alberts MJ, Hulette C, Crain B, Goldgaber D, Roses AD (1993) Association of apolipoprotein E allele E4 with late-onset familial and sporadic Alzheimer's disease Neurology 43 1467–1472

Strain SM, Tasker RAR (1991) Hippocampal damage produced by systemic injections of domoic acid in mice. Neuroscience 44 343–352

Strittmatter WJ, Saunders AM, Schmechel D, Pericak-Vance M, Enghild J, Salvesen GS, Roses AD (1993a) Apolipoprotein E high-avidity binding to β-amyloid and increased frequency of type 4 allele in late-onset familial Alzheimer disease Proc Natl Acad Sci USA 90: 1977–1981

Strittmatter WJ, Weisgraber KH, Huang DY, Dong LM, Salvesen GS, Pericak-Vance M, Schmechel D, Saunders AM, Goldgaber D, Roses AD (1993b) Binding of human apolipoprotein E to synthetic amyloid beta peptide isoform-specific effects and implications for late-onset Alzheimer disease Proc Natl Acad Sci USA 90 8098–8102

Strittmatter WJ, Weisgraber KH, Goedert M, Saunders AM, Huang D, Corder EH, Dong LM, Jakes R, Alberts MJ, Gilbert JR (1994) Hypothesis. microtubule instability and paired helical filament formation in the Alzheimer disease brain are related to apolipoprotein E genotype Exp Neurol 125· 163–171

Terry RD, Hansen L, Masliah E (1994) Structural alterations in Alzheimer disease In. Terry RD, Katzman R (eds) Alzheimer disease Raven Press, New York, pp 179–196

Urca G, Urca R (1990) Neurotoxic effects of excitatory amino acids in the mouse spinal cord quisqualate and kainate but not N-methyl-D-aspartate induced permanent neural damage Brain Res 529 7–15

Zhang SH, Reddick RL, Piedrahita JA, Maeda N (1992) Spontaneous hypercholesterolemia and arterial lesions in mice lacking apolipoprotein E Science 258 468–471

Very Low Density Lipoprotein Receptor and Alzheimer's Disease

*K. Ikeda**, *T. Aizawa*, *S. Haga*, *N. Otsuka*, *M. Mizuguchi*, *Y. Namba*, *R. Machinami*, *S.U. Kim*, and *T. Yamamoto*

Summary

The very low density lipoprotein receptor (VLDLR) is a recently discovered member of a protein family that includes the low density lipoprotein (LDL) receptor and the LDL receptor-related protein/α_2-macroglobulin receptor (LRP). The VLDLR specifically binds apolipoprotein E (apoE)-containing lipoproteins, including VLDL. Since antibodies to apoE immunostain senile plaques, neurofibrillary tangles and certain neuronal populations in autopsy brains from Alzheimer's disease patients, and since the message of this new apoE-binding receptor is abundant in the brain tissue, we examined the possibility that the receptor is involved in the pathogenesis of Alzheimer's disease, in which apoE has been implicated.

We first examined cell types that express the VLDLR in rat and human brain tissue by immunohistochemistry using monoclonal antibodies that recognize amino- and carboxyl-terminal portions of human and rat VLDLR, respectively. Immunoreactivity was found in most of the neurons throughout rat and human brains. Some astrocytes were also stained. We then examined primary cultured cells from human fetal brain. Intense staining was observed in most of the neurons. Glial fibrillary acidic protein (GFAP) positive astrocytes were weakly stained. To confirm the latter finding, pure astrocyte cultures were prepared by shaking off the loosely attached neurons and were examined by immunoblotting and reverse transcriptase (RT) polymerase chain reaction (PCR) methods. In immunoblotting experiments of cell extracts, the VLDLR antibodies identified a 130-kDa protein, whose apparent molecular weight was similar to that observed in extracts of cells transfected with the human VLDLR cDNA and tissue extracts of autopsied brain tissue. In RT-PCR analysis, total RNA from cultured astrocytes provided two PCR products, corresponding, respectively, to the portions of the VLDLR transcript (type I) and a variant transcript (type II) that lacks 84 nucleotides. These data demonstrated that the VLDLR was expressed both in neurons and astrocytes in brain and in culture.

---

*Department of Ultrastructure and Histochemistry, Tokyo Institute of Psychiatry, 2-1-8 Kamikitazawa, Setagaya, Tokyo 156, Japan

A D Roses et al (Eds)
Apolipoprotein E and Alzheimer's Disease

With these data in mind, we examined formalin-fixed, paraffin-embedded brain tissues (hippocampus and frontal lobe) from 18 Alzheimer's disease patients and 15 age-matched controls by a standard immunoperoxidase method using the VLDLR antibodies. In Alzheimer's disease cases, intense staining was observed in a large number of granule cell neurons and pyramidal neurons of hippocampus, with less intense staining in neurons of neocortex. Staining was restricted to neuronal perikarya and proximal processes, and its pattern was cytoplasmic and granular. Neuronal staining was weak in control brains, when compared to Alzheimer's disease brains. The VLDLR immunoreactivity was also seen in reactive astrocytes and dystrophic neurites in senile plaques. Senile plaques having the VLDLR immunoreactivity were stained with apoE antibody in the adjacent sections.

Taken together with a report of similar immunohistochemical findings for another apoE-binding lipoprotein receptor, the LRP, our data support a notion that apoE-binding lipoprotein receptors play roles in the formation of senile plaques. Since both apoE and βPP, a precursor protein that contains an amyloidogenic Aβ sequence, appear to be internalized through receptor-mediated endocytosis in neurons and processed in a specific, yet unidentified manner, it is of interest to speculate that common neuronal subcellular organelles might serve as sites where the two proteins can interact, both of which are implicated in AD and Aβ amyloidogenesis, respectively.

Introduction

Alzheimer's disease (AD) is characterized by synapse and neuron loss and protein abnormalities involving amyloid β-protein (Aβ) deposition and neurofibrillary tangles (De Kosky and Sheff 1990; Terry et al. 1991; Braak and Braak 1991; Hyman et al. 1993). A recent study by proton magnetic resonance spectroscopic imaging in AD patients revealed a decrease in signal intensities of *N*-acetylaspartate, suggesting axonal injury and membrane alterations in the brain (Meyerhoff et al. 1994). Prominent changes in membrane lipid composition and regional membrane instability have also been reported for postmortem brains of AD patients (Ellison et al. 1987; Nitsch et al. 1992; Ginsberg et al. 1993), and primary myelinopathy has been implicated in this disease condition (Englund et al. 1988). Thus, diffuse or regional membrane alterations involving synaptic and neuronal membrane, axons and myelin are probably involved in AD.

Apolipoproteins are lipid carrier molecules that play a crucial role in regulating the metabolism of lipid following peripheral nerves (for review, see Mahley 1988). Apolipoprotein E (apoE) in particular is unique among apolipoproteins in that it has a special relevance to nervous tissue. During development or after injury in the peripheral nervous system, apoE coordinates the mobilization and redistribution of cholesterol in repair, growth and maintenance of myelin and neuronal membranes (Ignatius et al. 1986; Boyles et al. 1989). In the central nervous system (CNS), however, little is known about the physiologic roles of apoE, which is mostly localized to astrocytes. Recently, apoE has been shown to play a pivotal role in the CNS in the mobilization and redistribution of cholesterol and phospholipid during membrane remodeling associated with synaptic plasticity (Poirier et al. 1991, 1993; Poirier 1994).

In examining possible abnormalities in lipid metabolism in AD brain by immunohistochemistry using antibody against apoE, we found a large number of immunopositive astrocytes that were abundant in AD brains (Namba et al. 1991). However, an unexpected and far more intriguing finding was the presence of apoE immunoreactivity in amyloid deposits of blood vessels and senile plaques and some neurofibrillary tangles. Subsequent studies confirmed the association of apoE with β-amyloid and also pointed out the association with other types of amyloid, including those found in non-neural tissues (Wisniewsky and Frangione 1992).

Three major isoforms (E2, E3, E4) of apoE occur in the human population. They differ by single amino acid changes. Consistent with the ε4 allele of human apoE alleles (ε2, ε3, ε4) as a risk factor for AD (Corder et al. 1993; Saunders et al. 1993; Poirier et al. 1993; Ueki et al. 1993), ApoE4 has been found to have a pronounced ability, compared to apoE3 and E2, to enhance the spontaneous in vitro fibrillogenesis of Aβ protein (Ma et al. 1994; Wisniewski et al. 1994), an amyloidogenic peptide that aggregates to form the primary component of senile plaques. With respect to the isotype-specific interaction of apoE with tau protein, whose hyperphosphorylated forms are the major constituents of neurofibrillary tangles, apoE3 but not E4 has been shown to bind tau with high avidity (Strittmatter et al. 1994a). This differential effect has led to the hypothesis that the apoE3 normally allows tau protein to stabilize microtubules and that its decrease or absence in patients with one or two apoE ε4 allele leads to a dissociation of tau from microtubules and its enhanced phosphorylation and polymerization into the pathological paired helical filaments (Strittmatter et al. 1994b).

Given that apoE appears to play some role in the pathogenesis of abnormalities involving both Aβ amyloid and neurofibrillary tangles, it was of particular interest to ask whether or not the specific receptors for apoE might be involved in the process. This question was first asked by Rebeck and Hyman and their colleagues (Rebeck et al. 1993), who examined AD brains by immunohistochemistry using antibodies to low density lipoprotein (LDL) receptor and the LDL receptor-related protein/α2-macroglobulin receptor (LRP), both of which are apoE-binding receptors. These workers found that anti-LRP antibodies immunostained senile plaques and reactive astrocytes in AD patients, in addition to the various populations of neurons that were equally stained in control brains. Together with their independent data suggesting that a mechanism is operating in human brain for clearing Aβ amyloid from the neuropil, they hypothesized that, in the process of eliminating amyloid by eliminating Aβ-apoE complexes through its internalization mechanism of receptor-ligand complexes into neurons and astrocytes, apoE4 may bring about decreased clearance, increased extracellular deposition or some other change in Aβ economy that augments the tissue burden of the peptide. A similar finding of LRP immunoreactivity in senile plaques was obtained by Tooyama et al. (1993), who pointed out that, in addition to the positive staining of senile plaques and neurofibrillary tangles, immunoreactivity of neurons was highly enhanced in AD patients compared to age-matched controls. Thus, LRP possessing the ability to bind with apoE has been implicated in the deposition of Aβ.

The VLDL receptor (VLDLR) is a recently discovered member of the LDL receptor family that currently includes the LDL receptor itself, the LDL receptor-related protein/α_2-macroglobulin receptor (LRP) and glycoprotein 330 (gp330; Takahashi et al. 1992; Sakai et al. 1994; Oka et al. 1994; Gåfvels et al. 1994; Webb et al. 1994). The VLDLR specifically binds apoE-containing lipoproteins, including VLDL, β-migrating VLDL (β-VLDL), and intermediate density lipoprotein (IDL), but not LDL (Takahashi et al. 1992). The VLDLR mRNA is expressed mainly in heart, skeletal muscle and adipose tissue, all of which utilize fatty acids as a source energy. Based on the ligand specificity and unique tissue distribution of the VLDLR, its physiologic function has been considered to provide extrahepatic energy-requiring tissues with fatty acids (Takahashi et al. 1992; Yamamoto et al. 1993). Since the mRNA of this new apoE-binding receptor is also abundant in brain tissue (Takahashi et al. 1992; Oka et al. 1994; Gåfvels et al. 1994; Webb et al. 1994), the VLDLR may be involved in the formation of AD-related abnormal structures, in which apoE has been implicated. We report here the results of immunohistochemical studies of the VLDLR in AD and control brains.

Results and Discussion

Immunohistochemistry Using Monoclonal Antibodies to the Very Low Density Lipoprotein Receptor

We prepared monoclonal antibodies (R2 and C4) to the VLDLR by immunizing mice with synthetic peptides corresponding, respectively, to amino-terminus (GRKTKCEASQFQCTNGR, 1–17 residues) and carboxyl-terminus (GHTYPAISVVSTDDDLA, 830–846 residues) of rabbit VLDLR (Takahashi et al. 1992). The amino-terminal sequence shows 79% and 78% homology with human and rat VLDLR, respectively (Sakai et al. 1994; Jokinen et al. 1994). The nine carboxyl residues of this peptide are identical among the three species. The carboxyl-terminal peptide is identical in sequence in the three species. Both monoclonal antibodies detected a 130-kDa protein (Fig. 1) in western blot analysis of membrane fraction from CHO-1 (ldlA-7) cells that were transfected with human VLDLR cDNA and pSV2-Neo (Sakai et al. 1994). Monoclonal antibody C4 did not detect the LDL

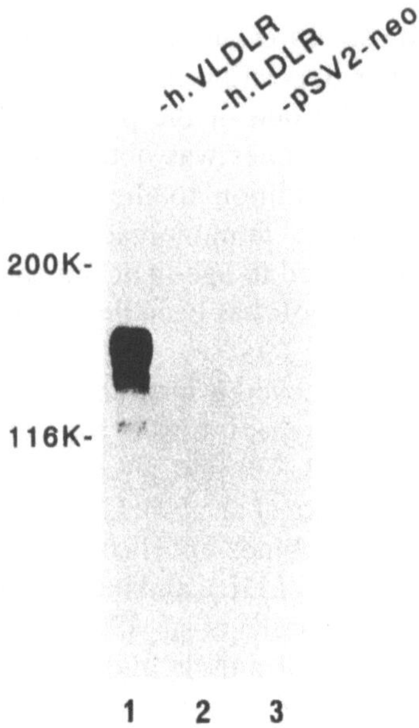

Fig. 1. Immunoblotting using monoclonal antibody (C4) to the carboxyl-terminal peptide of the very low density lipoprotein receptor (*VLDLR*) Cell extracts of ldlA-1 cells transfected with human VLDLR cDNA, human low-density lipoprotein receptor (*LDLR*) cDNA and pSV2-neo were applied to lanes *1*, *2*, and *3*, respectively Monoclonal antibody (C4) detected a 130-kDa protein

receptor that was expressed in cells transfected with human LDL receptor, whereas R2 weakly crossreacted with it.

Since the VLDLR mRNA is most abundant in heart, skeletal muscle and kidney, we examined formalin-fixed, paraffin-embedded sections from these tissues by the immunoperoxidase method using VLDLR monoclonal antibodies. In cardiac and skeletal muscles, immunoreactivity was observed in myoplasm in the form of coarse granules (Fig. 2). In the kidney, immunoreactivity was mainly localized in the epithelia of proximal convoluted tubules. Immunoreactivity was most prominent in the brush border and luminal portion of cells (Fig. 2). Interestingly, similar localization has been reported for apoE in this tissue (Lin et al., unpublished observation). Thus, apoE and its receptor, the VLDLR, may take part in the reabsorption process of the kidney, the function possibly unrelated to lipoprotein metabolism. Very little VLDLR mRNA is expressed in the liver (Takahashi et al. 1992), despite the importance of cholesterol homeostasis of this tissue. Consistent with this finding, immunohistochemistry showed no immunoreactivity in hepatocytes (Fig. 2). In rat tissues, similar localization of the VLDLR was observed (data not shown). Therefore, our antibodies to the VLDLR can detect both human and rat VLDLR in immunohistochemisty.

Very Low Density Lipoprotein Receptor Is Expressed in Neurons and Astrocytes in Rat and Human Brains

We first examined cell types that express the VLDLR in brain tissue. Preliminary studies have shown that the antigenicity of the receptor was sensitive to

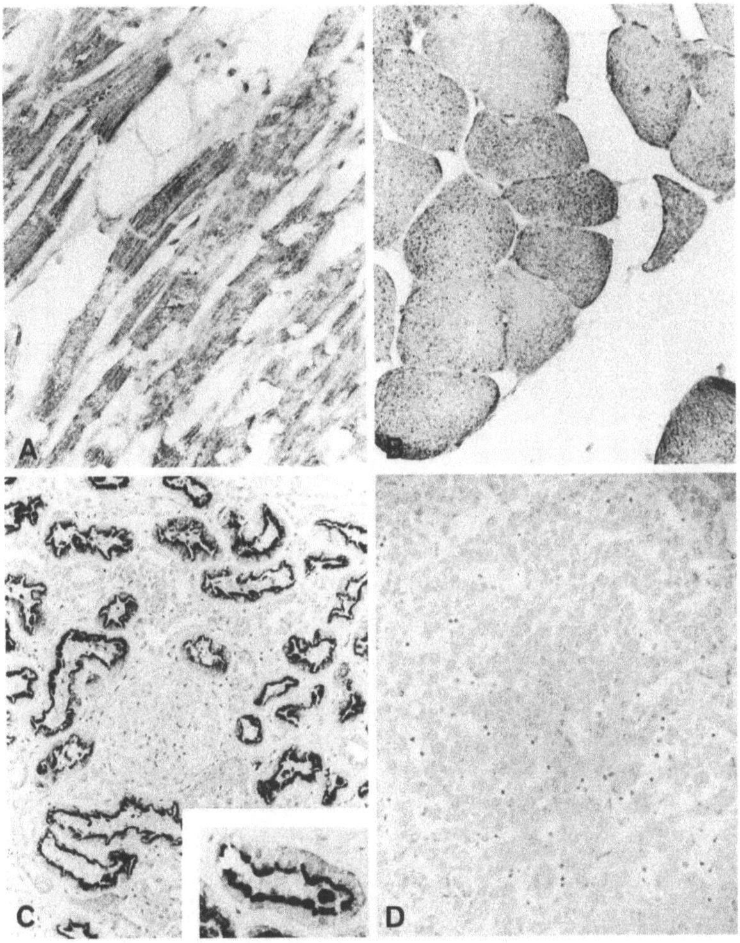

Fig. 2A–D. Immunostaining of human tissues for the very low density lipoprotein receptor (VLDLR) **A** Cardiac muscle **B** Skeletal muscle **C** Kidney **D** Liver Counterstained with hematoxylin **A,B,D** and *inset*, ×200, **C**, ×100 Immunoreactivity is seen in myoplasm in **A** and **B**. In **C**, immunoreactivity is intense in epithelial cells of proximal convoluted tubules. Brush border and luminal portion of the cells were strongly stained (*inset*) In **D**, there is no staining in hepatocytes

tissue fixation. Among several fixatives examined, Bouin solution was found to be the best for preserving the antigenicity. Rat brains were fixed in Bouin solution, embedded in paraffin and cut to thin sections. After deparaffinization, the sections were stained by immunoperoxidase technique using biotinylated second antibody-streptavidin (BSA) method. Antibodies to the VLDLR stained most of the neurons in variable intensities throughout the CNS (Fig. 3). Neurons in cerebellum, brain stem, hippocampus, thalamus, and spinal cord were strongly stained. Cortical neurons were less intensely stained;

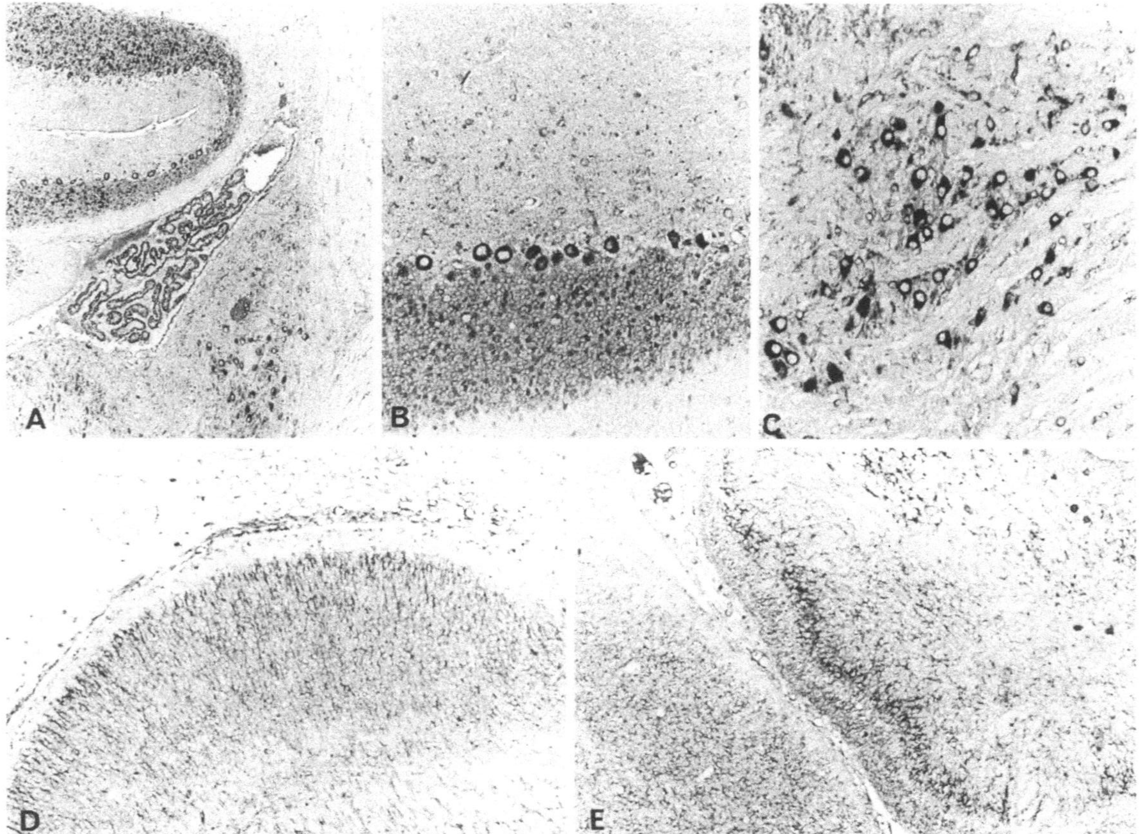

Fig. 3A-E. Immunostaining of rat brain for the very low density lipoprotein receptor (VLDLR) **A–C** Adult rat: cerebellum and pons (**A**, ×50), cerebellum (**B**, ×100), pons (**C**, ×200) **D,E** Fetal rat of 20 days in utero cerebrum (**D**, ×100), hippocampus (**E**, ×100) **A** A positive reaction is seen in Purkinje cells, cells in the granular layer, and neurons in the brain the pyramidal neurons showed relatively strong staining. Neuronal staining was restricted in perikarya and proximal processes, and the staining pattern was granular, suggesting its localization in the subcellular organelles. Axon and myelin were not stained. Immunoreactive glial cells were occasionally encountered. Double immunostaining using antibodies to the VLDLR and glial fibrillary acidic protein (GFAP) revealed that most of the VLDLR-positive cells were positive for GFAP, indicating localization in astrocytes. Among other tissue elements of the brain, choroid plexus epithelial cells, ependymal cells and smooth muscle of meningeal vessels were positively stained. Immunoreactivity was not found in capillary endothelial cells. In rat embryo (20 days), neurons in cerebral cortex and hippocampus were stained with these antibodies (Fig. 3). In the samples of human brain tissue that were fixed in Bouin solution, we observed a similar staining (Fig. 4). Therefore, the VLDLR is expressed in both neurons and astrocytes in rat and human CNS.

Very Low Density Lipoprotein Receptor Is Expressed in Neurons and Astrocytes in Primary Human Fetal Brain Cultures

To confirm the localization of the VLDLR in neurons and astrocytes, we then examined primary human brain cell cultures. Human fetal brains of 12–15 weeks, legally and therapeutically aborted, were dissected into small pieces and digested with trypsin. The dissociated cells were plated in culture flasks and incubated. After 2–4 weeks, cells consisted of a confluent basal layer of flat cells consisting of GFAP-positive astrocytes (Fig. 5C) and of a top layer of small bipolar cells consisting of MAP2-positive neurons (Fig. 5B). Using the immunofluorescence method, antibodies to the VLDLR strongly stained almost all bipolar neurons (Fig. 5D,E). Immunofluorescence was conspicuous in neurites as well perikarya. Varicosities were also positively stained. Cell surface distribution of the VLDLR was observed when viable cells were stained with R2 antibody, which recognizes an extracelluar epitope of the receptor.

In addition to the neuronal staining, weak fluorescence was consistently observed in astrocytes forming the basal layer (Fig. 5E). To confirm the expression of VLDLR in this cell type, we obtained astrocyte-enriched preparations from plastic surface-adherent cells and examined them by Western blot analysis. Antibodies to the VLDLR identified a 130-kDa protein in the blot analysis of cell extracts, whose apparent molecular weight was similar to that observed in tissue extracts of autopsied human brain tissue (Fig. 6).

Astrocytic localization of the receptor was further supported by reverse transcriptase (RT)-PCR analysis. Human VLDLR has been known to have two forms (type I and type II), type II lacking 28 amino acids in the *O*-linked sugar domain (Sakai et al. 1994). To detect DNA fragments of two forms in RT-PCR analysis, we selected a pair of oligonucleotide primers that flank the deletion of 84 nucleotides (a sense primer, oligo 2119: 5′-

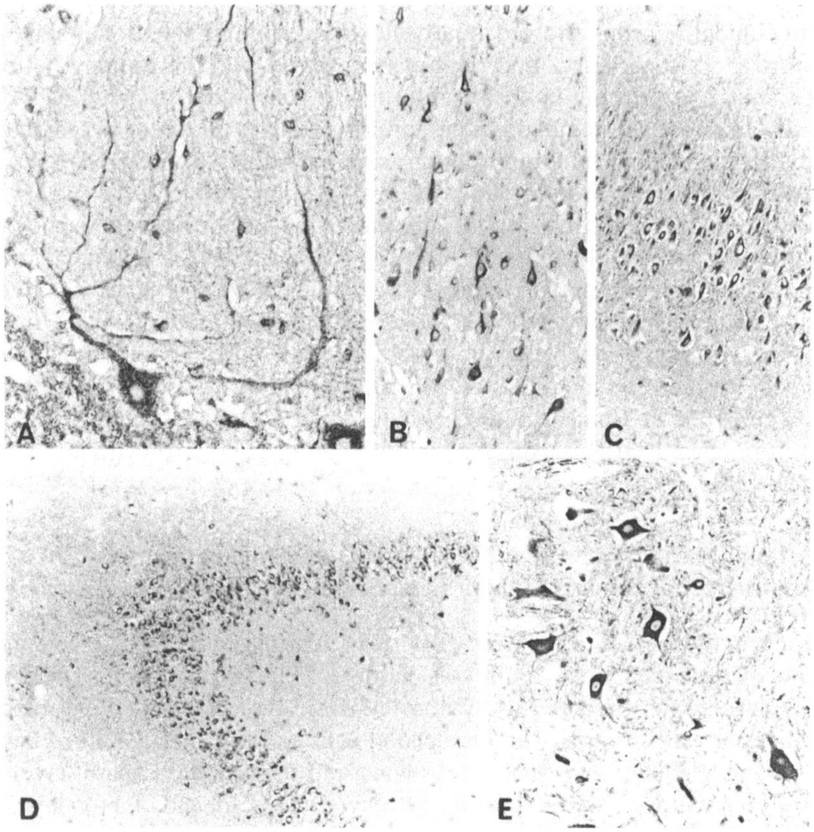

Fig. 4A–E. Immunostaining for the very low density lipoprotein receptor (VLDLR) of human control brain tissue fixed in Bouin solution **A** Cerebellum (×250) **B** Cortex (×100) **C** Pyramidal layer of hippocampus (×100). **D** Fascia dentata of hippocampus (×100) **E** Anterior horn of spinal cord (×200) Purkinje cells (**A**), cortical pyramidal neurons (**B**), hippocampal pyramidal neurons and granule neurons (**C,D**), and motor neurons of spinal cord (**E**) are storngly stained Glial cells in cerebellar molecular layer are also stained (**A**)

CAACCTGAATGATGCCCAAG-3′; an antisense primer, oligo 2442R, 5′-ACACTGACCTCTGATACTGC-3′). Total RNA from THP-1 cells used as a positive control gave two bands of 324 and 240 nucleotides (Fig. 7). When total RNA from primary cultured astrocytes or human brain tissue was examined, similar bands were identified in each sample. Type I and II differed in amounts among various cells in culture (data not shown). Thus, the VLDLR is expressed in both neurons and astrocytes in primary human fetal brain cultures.

Fig. 5A–E. Immunostaining of primary cultured human fetal brain cells **A** Phase contrast micrograph (×250) **B** Immunoperoxidase staining for microtubule-associated protein (MAP2, ×250) **C** Immunoperoxidase staining for glial fibrillary acidic protein (GFAP, ×250) **D,E** Immunofluorescence staining for the VLDLR (×300, ×600) Bipolar neurons on the bed layer of astrocytes are strongly positive Astrocytes show a weak immunostaining

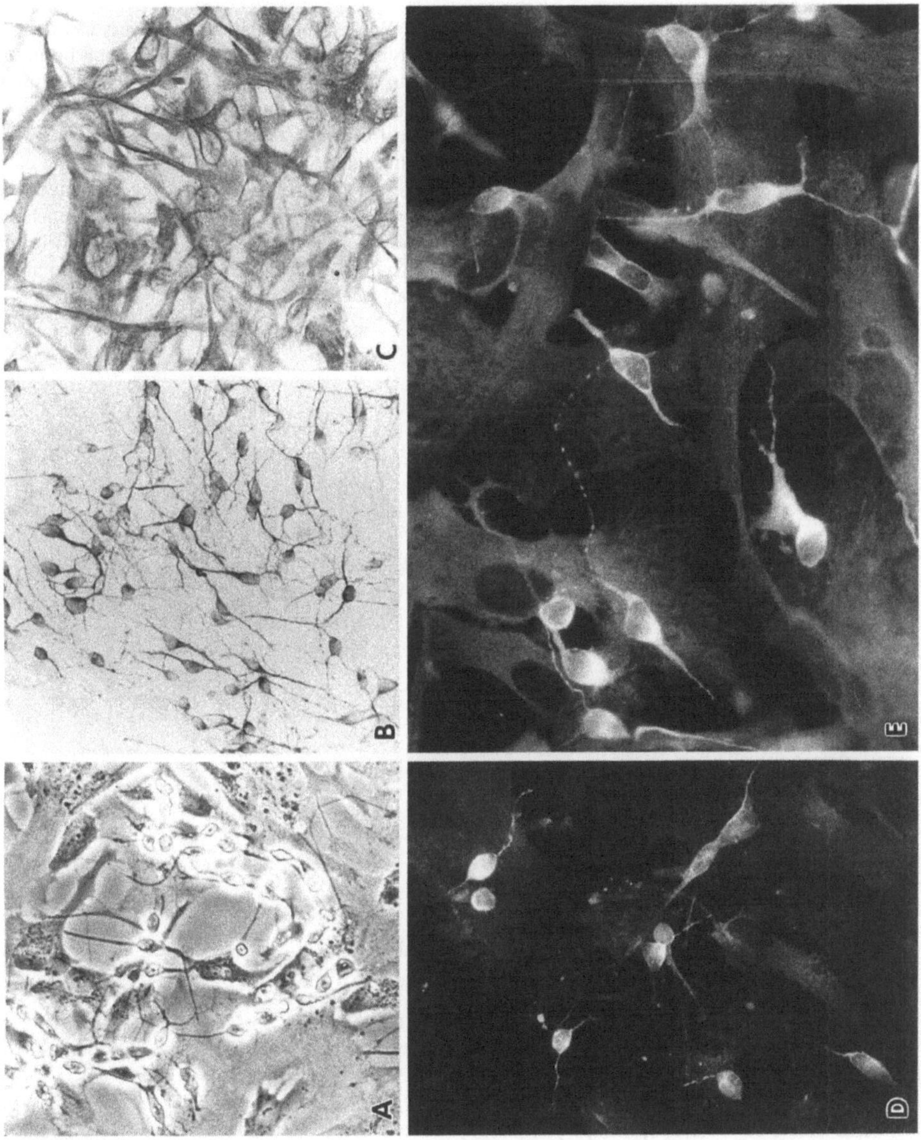
A
B
C
D
E

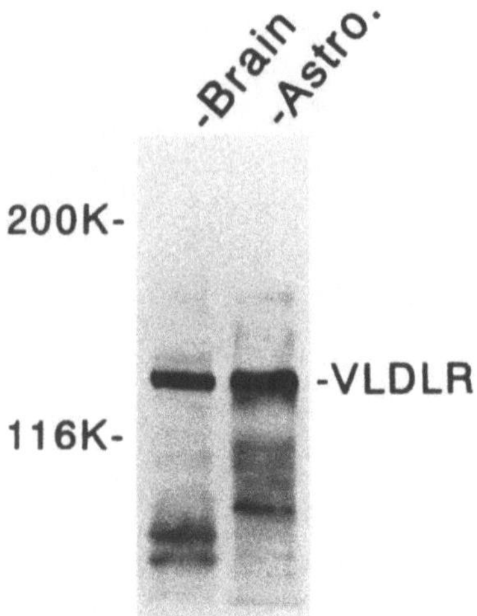

Fig. 6. Western blot analysis of the very low density lipoprotein receptor (VLDLR) in human brain tissue extracts and cell extracts of primary cultured human astrocytes Samples corresponding to 20 μg protein from the respective sources were resolved by sodium dodecyl sulfate polyacrylamide gel electrophoresis (SDS-PAGE) in 7 5% gel, blotted to a membrane, and immunostained using monoclonal antibody (C4) M_r markers are indicated on the *left* *Brain*, human brain, *Astro*, primary cultured human brain astrocyte

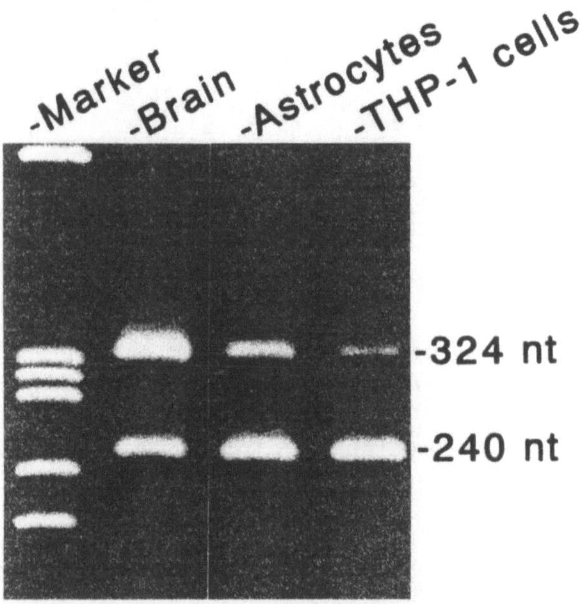

Fig. 7. Reverse transcriptase-polymerase chain reaction (RT-PCR) analysis of the very low density lipoprotein receptor (VLDLR) mRNA Total RNA (2 μg) from human brain tissue, primary cultured human brain astrocytes, or THP-1 human monocyte leukemia cells was used for this analysis. The resulting PCR products were separated on a polyacrylamide gel and stained with ethidium bromide *Hae*III-digested φ X174 RF I DNA was used as a molecular size marker *nt*, nucleotide

Very Low Density Lipoprotein Receptor Immunoreactivity in Brains of Patients with Alzheimer's Disease

We next studied the localization of the VLDLR in postmortem human brain tissue of AD patients and control subjects, using monoclonal antibodies to the receptor. Brains were obtained at autopsy from 18 AD patients (mean age, 71 years; range, 65–78) and 15 age-matched controls (mean age, 64 years; range, 62–80), fixed in formalin and embedded in paraffin. The areas examined were the hippocampus, entorhinal cortex, and occipitotemporal and frontal cortices. To unmask epitopes of the VLDLR in formalin-fixed, paraffin-embedded human brain sections, we checked several antigen retrieval methods, including protease digestion, formic acid treatment, a microwave procedure at acidic condition (Evers and Uylings 1994), and hydrated autoclaving (Shin et al. 1991). Since hydrated autoclaving substantially retrieved antigenicity of the VLDLR, we employed this procedure for the human autopsied materials.

In control brain tissue, the VLDLR immunoreactivity was weakly observed in hippocampal and neocortical neurons (Fig. 8B,D). In the hippocampus, granule neurons in the dentate gyrus and pyramidal neurons in CA4,

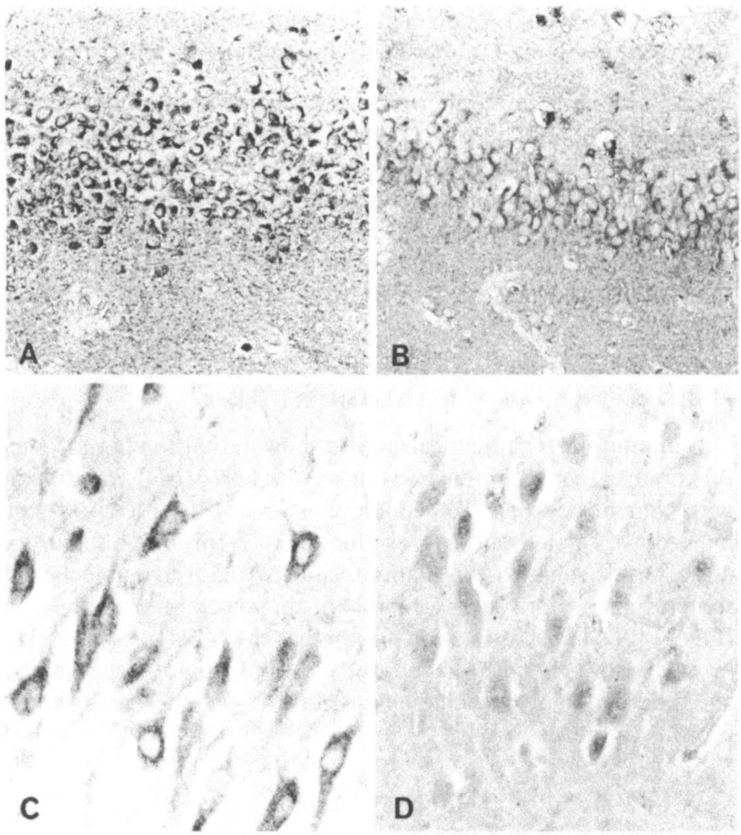

Fig. 8A–D. Immunostaining of Alzheimer's disease (AD) and control brain sections for the very low density lipoprotein receptor (VLDLR) **A,C** AD hippocampus **B,D** Control hippocampus **A,B** Granule cells of fascia dentata (×125) **C,D** Pyramidal neurons (×250) Immunoreactivity is more enhanced in **A** and **B** than in **B** and **D**, respectively

CA3, CA1 hippocampal and subicular subfields were stained. In occipitoparietal and frontal cortices, positively stained pyramidal neurons in layer III were discernible. Glial staining was not conspicuous in both gray and white matter.

In AD brain tissue, immunoreactivity was significantly enhanced, showing a clear staining of dentate gyrus and pyramidal layer of hippocampus. In hippocampus, immunoreactivity was intensely observed in granule cells and pyramidal neurons (Fig. 9). Neurofibrillary tangles were not associated with immunoreactivity. When staining intensity was compared between AD and normal brains for granule cells of the dentate gyrus and pyramidal neurons in CA1, both neuronal populations showed stronger staining in AD brain (Fig. 9). The difference in staining intensity of cortical neurons between AD patients and controls was not conspicuous. Reactive astrocytes were also strongly stained.

Very Low Density Lipoprotein Receptor Immunoreactivity Is Present in Senile Plaques in Brains of Patients with Alzheimer's Disease

There were a large number of senile plaques in the dentate gyrus molecular layer and neocortex in all AD brain tissues. Antibodies to the VLDLR stained most of the senile plaques in hippocampus that were also positive for antibody to apoE in adjacent sections (Fig. 10A,B). Closer inspection revealed that the VLDLR-immunoreactivity in senile plaques was mainly restricted to dystrophic neurites. Consistent with this finding, antibodies to the VLDLR did not stain diffuse plaques that lack dystrophic neurites in neocorex (Fig. 10E,F).

Is the Very Low Density Lipoprotein Receptor Expression Enhanced in Brains of Patients with Alzheimer's Disease?

Our immunohistochmemical studies showed that the hippocampal neurons in AD brains are more intensely stained with anti-VLDLR antibodies, compared with control neurons. To see whether or not the receptor expression is enhanced, we carried out Western blot analysis of temporal cortex tissue taken at autopsy from two AD patients and two elderly subjects. Two AD brain samples gave more intense bands of the receptor than the control samples (Fig. 11). To substantiate the notion that the VLDLR expression is enhanced in AD brain neurons, further studies, employing a large number of samples and other techniques, including quantitative RT-PCR assay and in situ hybridization, are necessary.

Very Low Density Lipoprotein Receptor Expression in Neuronal Cells in Culture

Our studies suggested the possibility that the VLDLR is upregulated in neurons in AD patients. The 5′-flanking region of the human VLDLR gene has been known to contain two copies of potential sterol regulatory element-1

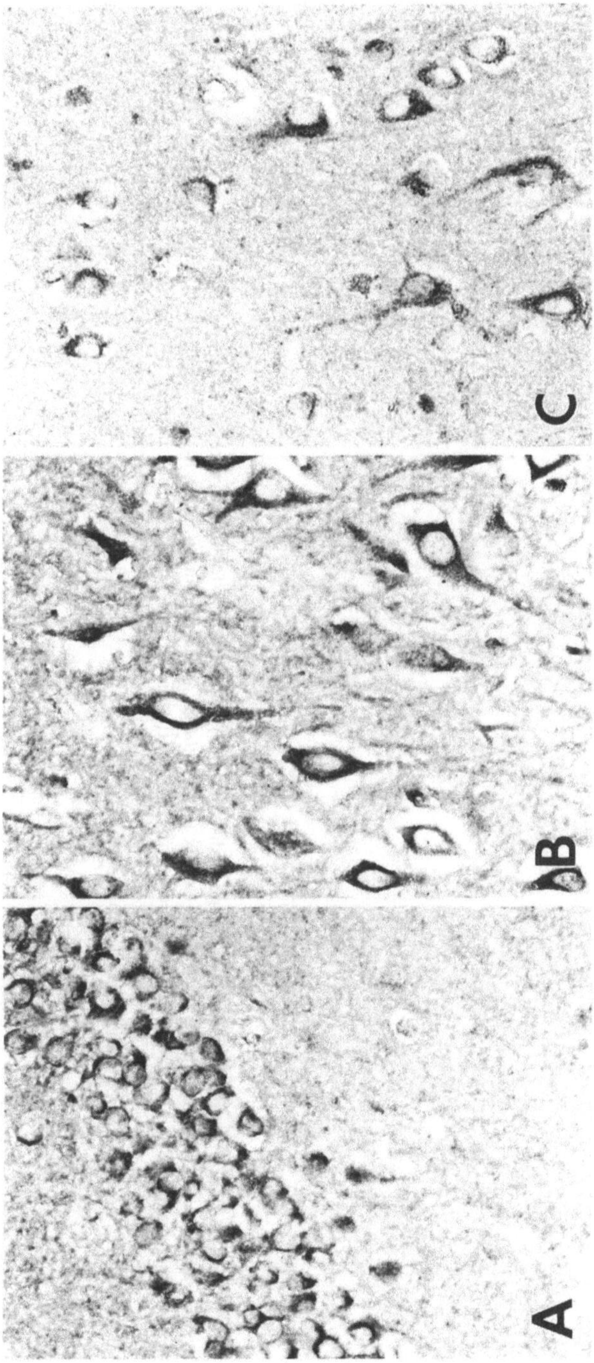

Fig. 9A–C. Immunostaining of Alzheimer's disease (AD) brain tissue for the very low density lipoprotein receptor (VLDLR). Strong imunoreactivity is seen in granule cells (**A**), pyramidal neurons (**B**), and cortical pyramidal neurons (**C**) Magnification, ×400

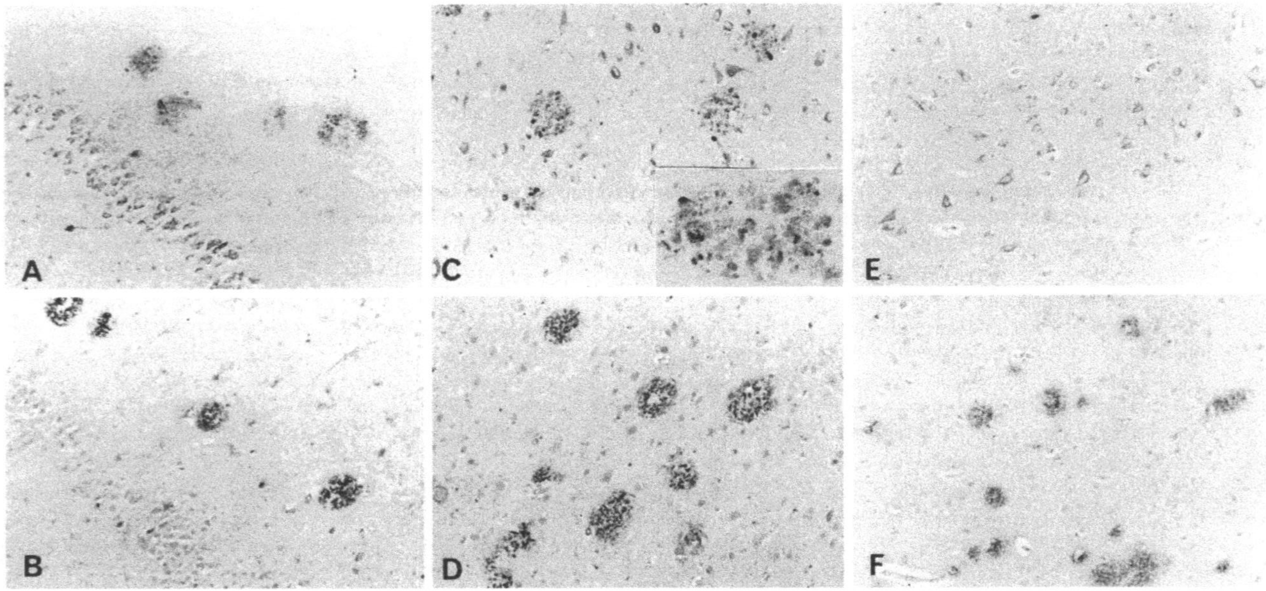

Fig. 10A–F. Immunostaining of Alzheimer's disease (AD) brain sections for the very low density lipoprotein receptor (VLDLR) and apoE **A–D** Hippocampus (×100) **E,F** Cortex (×100) *Inset*, ×250 Immunostaining for the VLDLR (**A,C,E**) and apoE (**B,D,F**) In hippocampus, senile plaques are positively stained by antibodies to the VLDLR Immunoreactivity is mainly confined to dystrophic neurites (*inset*) VLDLR-immunoreactive senile plaques are positive for apoE immunostaining in adjacent sections Diffuse plaques in cortex are not stained with antibodies to the VLDLR (**E,F**)

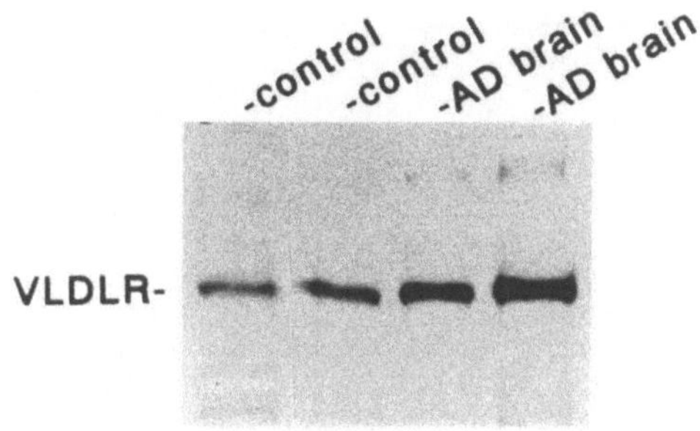

Fig. 11. Western blot analysis of the very low density lipoprotein receptor (*VLDLR*) in Alzheimer's disease (*AD*) and control brain tissues Immunopositive protein for the VLDLR is somewhat intenser in staining in AD samples than in control samples

(SRE-1) and a number of half sites for interaction with glucocorticoid, androgen, and estrogen receptors (Sakai et al. 1994). The SRE-1 is functional in the LDL receptor gene, mediating down-regulation of the LDL receptor by sterols. However, the levels of the VLDLR mRNA are unchanged in the presence of LDL or β-VLDL in THP-1 cells (Sakai et al. 1994). In contrast, the VLDLR mRNA levels were very recently found to be increased in rabbit cardiac muscle when estradiol was systemically administrated into the animal (Masuzaki et al. 1994).

We examined the possible change in expression of the receptor in cultured neurons following serum deprivation of the culture medium. Two human neuroblastoma cell lines, IMR32 (Tumilowicz et al. 1970) and NB-1 (Imashuku et al. 1973), were used in this experiment. In both cell lines, a few cells showed immunoreactivity to antibodies to the VLDLR. Total RNA from the cell lines gave two DNA fragments in RT-PCR assay, as described above. Three and five days after the cultivation of cells in serum-free medium, the cells produced extensive neurites that were strongly stained by the immunofluorescence method using antibodies to the VLDLR (Fig. 12). RT-PCR assay of the cells harvested at these points revealed an increase in the amount of the VLDLR mRNA (data not shown). Thus, the levels of the VLDLR mRNA can be altered by culture condition.

Are ApoE Receptors Involved in the Pathogenesis of Alzheimer's Disease?

We demonstrated that the newly found apoE-binding receptor, the VLDLR, is expressed in human brain neurons throughout the brain. Several groups have obtained similar neuronal localization for another apoE-binding receptor, LRP (Moestrup et al. 1992; Wolf et al. 1992; Rebeck et al. 1993; Tooyama et al. 1993). Thus, among the LDL receptor family members that can interact with apoE-enriched lipoprotein molecules, both LRP and VLDLR are expressed in the neurons of CNS. The LDL receptor, a prototype of the lipoprotein receptors which binds to both apoB and apoE, is also expressed in the

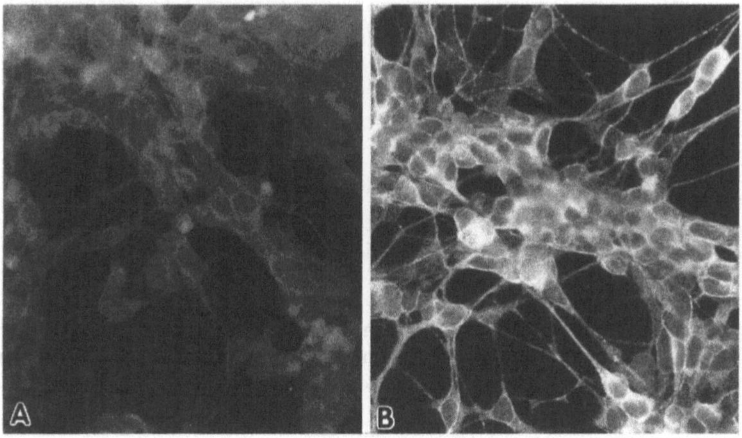

Fig. 12A,B. Immunofluorescence staining of NB1 human neuroblastoma cells for the very low density lipoprotein receptor (VLDLR) A few cells are positively stained for the VLDLR in control cultures (**A**) A large number of cells become immunopositive for the VLDLR when cultured in serum-free medium for 3 days (**B**)

brain neurons as shown by in situ hybridization (Swanson et al. 1988); however, the neuronal localization is not always possible by immunohistochemistry (Pitas et al. 1987; Rebeck et al. 1993), presumably due to low antigen density.

LRP is suggested to be functional in the CNS neurons, based on the fact that cultured cortical neurons bind with the 39-kDa receptor-associated protein that is a ligand recognized by LRP (Bu et al. 1994). This finding may suggest that the VLDLR is functional in the CNS neurons, because the VLDLR is expressed in brain neurons in situ and in vitro, as demonstrated here, and because this receptor is known to bind with the 39-kDa receptor-associated protein (Battey et al. 1994; Simonsen et al. 1994). By immunohistochemistry, both LRP and the VLDLR are localized on neuronal cell bodies and proximal processes, whereas the LDL receptor appears to be localized in neuropil (Rebeck et al. 1993). Their roles in the brain may differ in lipoprotein and protein metabolism.

In AD brain tissue, antibodies to the VLDLR stained senile plaques, with immunoreactivity mostly confined to dystrophic neurites. When the sections stained with antibodies to the receptor were compared with the adjacent sections that were stained with antibody to ApoE, the VLDLR-immunoreactive plaques were also positive for apoE. Colocalization of LRP and apoE has been reported in senile plaques (Rebeck et al. 1993). Therefore, two receptors capable of binding ApoE-enriched lipoproteins are localized in senile plaques, which are also associated with apoE, suggesting their roles in the formation of senile plaques.

The evidence that apoE immunoreactivity is associated with senile plaques (Namba et al. 1991; Wisniewski and Frangione 1992) and apoE binds senile plaques in situ (Richey et al. 1995) and Aβ peptide in vitro (Strittmatter et al. 1993; Wisniewski et al. 1993) supports the hypothesis that a direct interaction between apoE and Aβ is crucial in the AD pathogenesis. Aβ-apoE complexes are found in cerebrospinal fluid. ApoE could help the complexes bind to the proteogylcans that are abundantly found in cell surface, neuropil and senile plaques (Snow et al. 1989; Young et al. 1989), by means of its capacity to bind with heparan sulfate proteoglycans (Mahley et al. 1979; Ji et al. 1994). Our computer survey detected two sequences in human apoE, RVRLASHLR (134–142) and RKLRKRLLR (142–150), both of which fit in a common hyaluronate binding motif, B(X7)B, in which B is either R or K and X7 contains no acidic residues and at least one basic amino acid (Yang et al. 1994). Thus, accumulation of Aβ peptide on proteoglycans mediated by apoE may contribute to the concentration of Aβ, favoring the amyloid fibril formation.

Aβ-apoE complexes may be taken up by cells that express apoE-binding receptors, including the LDL receptor, LRP and the VLDLR. Based on the presence of LRP in neurons and reactive astrocytes, Rebeck et al. hypothesized that apoE-Aβ complexes may be taken up by LRP and therefore cleared from the neuropil. They suggest that the apoE-E4 isoform may have a decreased ability to clear Aβ from the neuropil, due to altered apoE-Aβ interactions, or due to lower levels of apoE in individuals possessing the apoE-E4 phenotype, bringing about the higher levels of Aβ in the neuropil and the subsequent possibility of amyloid fibril formation. Since the VLDLR is also present in neurons and astrocytes in human brain, this receptor could play a similar role in this scheme.

Our data suggest that the VLDLR may be produced more in neurons of patients with AD than in those of controls. Interestingly, expression of LRP is reported to be significantly enhanced in neurons of AD (Tooyama et al. 1993). Taken together, the apoE receptor-mediated endocytosis may be more active in AD brain neurons. If this is the case, then ligands including apoE-enriched lipoproteins, apoE-Aβ complexes and other molecules may be taken up by neurons and astrocytes more excessively in AD. This may be consistent with the finding that apoE immunoreactivity is detected in hippocampal and cortical neurons – cells that do not synthesize apoE – in most of the AD patients and in occasional aged control subjects (Strittmatter et al. 1994b; Han et al. 1994). Increased immunostaining of cerebral neurons has also been demonstrated in rabbits that were fed a cholesterol diet (Sparks et al. 1995). This phenomenon could be best explained by the receptor-mediated internalization of apoE-enriched lipoprotein by LRP and/or VLDLR, both of which are present in neurons and, unlike the LDL receptor, are not down-regulated by excessive amounts of cholesterol.

In cultured nonneuronal cells or hepatocytes in vivo, receptor-mediated endocytosis is well documented for the LDL receptor (Brown and Goldstein 1986) and LRP (Herz et al. 1990, 1992). This mechanism is also suggested for the VLDLR (Takahashi et al. 1992). Ligand-receptor complexes are internalized in coated pits and transported to endosomes where, in the acidic environment, the ligand is dissociated from its receptor, carried forward to the lysosome and degraded, while the receptor is recycled from the endosome to cell surface through a recycling vesicle. Neurons in the CNS might differ in intracellular processing of apoE, in view of the common knowledge that apolipoproteins are readily degraded in the cells after being internalized. Whether the presence of apoE in neurons detected by immunohistochemistry in AD patients and some aged controls is due to the enhanced uptake of apoE, caused by the excessive expression of apoE-binding lipoprotein receptors, or to a difference in intracellular processing of apoE in impaired neuronal cells or both remains to be elucidated.

Aβ peptide has been shown to be a product of normal cellular metabolism (Estus et al. 1992; Haass et al. 1992a; Shoji et al. 1992). Aβ peptide is derived by two pathways from the β-amyloid precursor protein (βPP). In addition to the secretory pathway (Esch et al. 1990; Sisoda et al. 1990), βPP is also processed and degraded within endosomal/lysosomal pathway where carboxyl-terminal fragments containing the entire Aβ sequence have been detected (Golde et al. 1992; Haass et al. 1992b). Recent studies have implicated the internalization of cell surface βPP in cultured cells as an important contributor to Aβ subsequently released into the medium (Koo and Squazzo 1994). In cultured neuronal cells, βPP internalized at growth cones has recently been shown to be transported retrogradely, although the organelle that transports it is not yet identified (Yamazaki et al. 1995). In neurons, which are the cells that have long been hypothesized to be the principal source of Aβ peptide in brain, both apoE and βPP appear to be internalized through receptor-mediated endocytosis and processed in a specific, yet unidentified manner. If the proteins share organelles, then one can expect the various interactions, as observed in vitro, between apoE and Aβ or carboxyl-terminal fragments of βPP containing the entire Aβ sequence. Neuronal subcellular organelles might serve as the sites where the two proteins that are implicated in AD and Aβ amyloidogenesis, respectively, can interact.

Acknowledgments. Grant support was provided by the Research Grant for Nervous and Mental Health Disorders (6A-2, KI) and the Health Science Research Grants (94A240400, KI) from the Ministry of Health and Welfare Japan and by the Special Research Grant for Alzheimer's Disease from the Tokyo Metropolitan Government (KI).

References

Battey FD, Gåfvels ME, FitzGerald DJ, Argraves WS, Chappell DA, Strauss III JF, Strickland DK (1994) The 39-kDa receptor-associated protein regulates ligand binding by the very low density lipoprotein receptor J Biol Chem 269: 23268–23273

Boyles JK, Zoellner CD, Anderson LJ, Kosik LM, Pitas RE, Weisgraber KH, Hui DY, Mahley RW, Gabicke-Haerter PJ, Ignatius MJ, Shooter EM (1989) A role for apolipoprotein E, apolipoprotein A-I, and low density lipoprotein receptors in cholesterol transport during regeneration and remyelination of the rat sciatic nerve J Cin Invest 83 1015–1031

Braak H, Braak E (1991) Neuropathological stageing of Alzheimer-related changes. Acta Neuropathol 82· 239–259

Brown MS, Goldstein JL (1986) A receptor-mediated pathway for cholesterol homeostasis Science 232 34–37

Bu G, Maksymovitch ES, Nerbonne JM, Schwartz AL (1994) Expression and function of the low density lipoprotein receptor-related protein (LRP) in mammalian central nervous system. J Biol Chem 269: 18521–18528

Corder EH, Saunders AM, Strittmatter WJ, Schmechel DE, Gaskel PC, Small GW, Roses AD, Haines JL, Pericak-Vance MA (1993) Gene dose of Apolipoprotein E type 4 allele and the risk of Alzheimer's disease in late onset families Science 261 921–923

De Kosky ST, Scheff SW (1990) Synaptic loss in frontal cortex biopsies in Alzheimer's disease correlation with cognitive severity Ann Neurol 27 457–464

Englund E, Brun A, Alling C (1988) White matter change in dementia of Alzheimer type Brain 111 1425–1439

Ellison DW, Beal MF, Martin JB (1987) Phosphoethanolamine and ethanolamine are decreased in Alzheimer's disease and Huntington's disease Brain Res 417 389–392

Esch FS, Keim PS, Beattie EC, Blacher RW, Culwell AR, Oltersdorf T, McClure D, Ward PJ (1990) Cleavage of amyloid β-sheet during constitutive processing of its precursor. Science 248 1122–1124

Estus S, Golde TE, Kunishita T, Blades D, Lowery D, Eisen M, Usiak M, Qu X, Tabira T, Greenberg BD, Younkin S (1992) Potentially amyloidogenic carboxyl-termial derivatives of the amyloid protein precursor Science 255 726–728

Evers P, Uylings HBM (1994) Microwave-stimulated antigen retrieval is pH and temperature dependent. J Histochem Cytochem 42 1555–1563

Gåfvels ME, Paavola LG, Boyd CO, Nolan PM, Wittmaack F, Chawla A, Lazar MA, Bucan LM, Angelin B, Straus III JF (1994) Cloning of a complementary deoxyribonucleic acid encoding the murine homolog of the very low density lipoprotein/apolipoprotein-E receptor expression pattern and assignment of the gene to mouse chromosome 19. Endocrinology 135 387–394

Ginsberg L, Atack JR, Rapoport SI, Gershfeld NL (1993) Regional specificity of membrane instablity in Alzheimer's disease brain. Brain Res 615 355–357

Golde TE, Estus S, Younkin LH, Selkoe DJ, Younkin SG (1992) Processing of the amyloid protein precursor to potentially amyloidogenic carboxyl-teminal derivatives. Science 255· 728–730

Haass C, Schlossmacher MG, Hung AY, Vigo-Pelfrey C, Mellon A, Ostaszewski BL, Lieberburg I, Koo EH, Schenk D, Teplow DB, Selkoe DJ (1992a) Amyloid β-peptide is produced by cultured cells during normal metabolism Nature 359. 322–325

Haass C, Koo EH, Mellon A, Hung AY, Selkoe DJ (1992b) Targeting of cell-surface β-amyloid precursor protein to lysosomes alternative processing into amyloid-bearing fragments Nature 357. 500–503

Han S-H, Hulette C, Saunders AM, Einstein G, Pericak-Vance M, Strittmatter WJ, Roses AD, Schmechel DE (1994) Apolipoprotein E is present in hippocampal neurons without neurofibrillary tangles in Alzheimer's disease and in age-matched controls. Exp Neurol 128· 13–26

Herz J, Kowal RC, Ho YK, Brown MS, Goldstein JL (1990) Low density lipoprotein receptor-related protein mediates endocytosis of monoclonal antibodies in cultured cells and rabbit liver. J Biol Chem 265 21355–21362

Herz J, Clouthier DE, Hammer RE (1992) LDL receptor-related protein internalizes and degrades uPA-PAI complexes and is essential for embryo implantation Cell 71: 411–421

Hyman BT, Marzloff K, Arigada PV (1993) The lack of accumulation of senile plaques or amyloid burden in Alzheimer's disease suggests a dynamic balance between amyloid deposition and resolution J Neuropathol Exp Neurol 52 594–600
Ignatius MJ, Gebicke-Haerter PJ, Skene JH, Schilling JW, Weisgraber KH, Mahley RW, Shooter EM (1986) Expression of apolipoprotein E during nerve degeneration and regeneration Proc Natl Acad Sci USA 83 1125–1129
Ignatius MJ, Shooter EM, Pitas RE, Mahley RW (1987) Lipoprotein uptake by neuronal growth cones in vitro Science 236 959–962
Imashuku S, Inui A, Nakamura T, Tanaka J, Miyake S (1973) Catecholamine metabolism in tissue culture cells of a neuroblastoma J Clin Endocrinol Metab 36 931–936
Ji Z-S, Fazio S, Lee Y-L, Mahley RW (1994) Secretion-capture role for apolipoprotein E in remnant lipoprotein metabolism involving cell surface heparan sulfate proteoglycans J Biol Chem 269 2764–2772
Jokinen EV, Landschulz KT, Wyne KL, Ho YK, Frykman PK, Hobbs HH (1994) Regulation of the very low density lipoprotein receptor by thyroid hormone in rat skeletal muscle J Biol Chem 269: 26411–26418
Koo EH, Squazzo SL (1994) Evidence that production and release of amyloid β-protein involves the endocytic pathway J Biol Chem 269 17386–17389
Ma J, Yee A, Brewer Jr B, Das S, Potter H (1994) Amyloid-associated proteins α 1-antichymotrypsin and apolipoprotein E promote assembly of Alzheimer β-protein into filaments Nature 372 92–94
Mahley RW (1988) Apolipoprotein E cholesterol transport protein with expanding role in cell biology Science 240 622–630
Mahley RW, Weisgraber KH, Innerarity TL (1979) Interaction of plasma lipoproteins containing apolipoprotein B and E with heparin and cell surface receptors Biochem Biophys Acta 575 81–91
Masuzaki H, Jingami H, Yamamoto T, Nakao K (1994) Effects of estradiol on very low density lipoprotein receptor mRNA levels in rabbit heart FEBS Lett 347 211–214
Meyerhoff DL, MacKay S, Constans J-M, Norman D, Van Dyke C, Fein G, Weiner MW (1994) Axonal injury and membrane alterations in Alzheimer's disease suggested by in vitro proton magnetic resonance spectroscopic imaging Ann Neurol 36 40–47
Moestrup SK, Gliemann J, Pallesen G (1992) Distribution of the α2-macroglobulin receptor/low density lipoprotein receptor-related protein in human tissue Cell Tissue Res 269 375–382
Namba Y, Tomonaga M, Kawasaki H, Otomo E, Ikeda K (1991) Apolipoprotein E immunoreactivity in cerebral amyloid deposits and neurofibrillary tangles in Alzheimer's disease and kuru plaque amyloid in Creutzfeldt-Jakob disease Brain Res 541 163–166
Nitsch RM, Blusztajn JK, Pittas AG, Slack BE, Growdon JH, Wurtman RJ (1992) Evidence for a membrane defect in Alzheimer disease brain Proc Natl Acad Sci USA 89: 1671–1675
Oka K, Ishimura-Oka K, Chu M-J, Sullivan M, Krushkal J, Li W-H, Chan L (1994) Mouse very-low-density-lipoprotein receptor (VLDLR) cDNA cloning, tissue-specific expression and evolutionary relationship with the low-density-lipoprotein receptor. Eur J Biochem 224: 975–982
Pitas RE, Boyles JK, Lee SH, Hui D, Weisgraber KH (1987) Lipoproteins and their receptors in the central nervous system J Biol Chem 262 14352–14360
Poirier J (1993) Apolipoprotein E polymorphysim and Alzheimer's disease Lancet 342 697–699
Poirier J (1994) Apolipoprotein E in animal models of CNS injury and in Alzheimer's disease Trends Neurosc 17 525–530
Poirier J, Hess M, May PC, Finch CE (1991) Astrocytic apolipoprotein E mRNA and GFAP mRNA in hippocampus after entorhinal cortex lesioning. Mol Brain Res 11. 97–106
Poirier J, Baccichet A, Dea D, Gauthier S (1993) Cholesterol synthesis and lipoprotein reuptake during synaptic remodeling in hippocampus in adult rats. Neuroscience 55: 81–90
Rebeck GW, Reiter JS, Strickland DK, Hyman BT (1993) Apolipoprotein E in sporadic Alzheimer's disease. allelic variation and receptor interactions Neuron 11 575–580

Richey PL, Siedlak SL, Smith MA, Perry G (1995) Apolipoprotein E interaction with the neurofibrillary tangles and senile plaques in Alzheimer disease. implications for disease pathogenesis Biochem Biophys Res Commun 208 657–663

Sakai J, Hishino A, Takahashi S, Miura Y, Ishii H, Suzuki H, Kawarabayasi Y, Yamamoto T (1994) Structure, chromosome location, and exchange of the human very low density lipoprotein receptor gene. J Biol Chem 269. 2173–2182

Saunders AM, Stittmatter WJ, Schmechel D, St George-Hyslop PH, Pericak-Vance MA, Joo SH, Rosi BL, Gusella JF, Crapper-MacLachlan DR, Alberts MJ, Hulette C, Crain B, Goldgaber D, Roses AD (1993) Association of apolipoprotein E allele ε4 with late-onset familial and sporadic Alzheimer's disease Neurology 43. 1467–1472

Shin R-W, Iwaki T, Kitamoto T, Tateishi J (1991) Hydrated autoclave pretreatment enhances tau immunoreactivity in formalin-fixed normal and Alzheimer's disease brain tissues Lab Invest 64: 693–702

Shoji M, Golde TE, Ghiso J, Cheung TT, Estus S, Shaffer LM, Cai X, McKay DM, Tintner R, Frangione B, Younkin SG (1992) Production of the Alzheimer amyloid β protein by normal proteolytic processing Science 258 126–129

Simonsen ACW, Heegaard CW, Rasmussen LK, Ellgaard L, Kjoller L, Christensen A, Etzerodt M, Andreasen P (1994) Very low density lipoprotein receptor from mammary gland and mammary epithelial cell lines binds and mediates endocytosis of Mr 40,000 receptor associated protein FEBS Lett 354 279–283

Sisoda SS, Koo EH, Beyreuther K, Unterbeck A, Price DL (1990) Evidence that β-amyloid protein in Alzheimer's disease is not derived by normal processing Science 248. 492–495

Snow AD, Lara S, Nochlin D, Wight TN (1989) Cationic dyes reveal proteoglycans structurally integrated within the characteristic lesions of Alzheimer's disease Acta Neuropathol 78 113–123

Sparks DL, Liu H, Gross DR, Scheff SW (1995) Increased density of cortical apolipoprotein E immunoreactive neurons in rabbit brain after dietary administration of cholesterol Neurosci Lett 187: 142–144

Strittmatter WJ, Weisgraber KH, Huang D, Dong L-M, Salvesen GS, Pericak-Vance M, Schmechel D, Saunders AM, Goldgaber D, Roses AD (1993) Binding of human apolipoprotein E to synthetic amyloid β peptide· isoform specific effects and implications for late-onset Alzheimer disease. Proc Natl Acad Sci USA 90· 8098–8102

Strittmatter WJ, Saunders A, Goedert M, Weisgraber KH, Dong L-M, Jakes R, Huang DY, Pericak-Vance M, Schmechel D (1994a) Isoform-specific interactions of apolipoprotein E with microtubule-associated protein tau implications for Alzheimer disease Proc Natl Acad Aci USA 91 11183–11186

Strittmatter WJ, Weisgraber KH, Goedert M, Saunders AM, Huang D, Corder EH, Dong L-M, Jakes R, Alberts MJ, Gilbert JR, Han S-H, Hulette C, Einstein G, Schmechel DE, Pericak-Vance MA, Roses AD (1994b) Hypothesis. microtubule instability and paired helical filament formation in the Alzheimer disease brain are related to apolipoprotein E genotype Exp Neurol 125· 163–171

Swanson LW, Simmons DM, Hofmann SL, Goldstein JL, Brown MS (1988) Localization of mRNA for low density lipoprotein receptor and a cholesterol enzyme in rabbit nervous system by in situ hybridization Proc Natl Acad Sci USA 85. 9821–9825

Takahashi S, Kawarabayashi Y, Nakai T, Sakai J, Yamamoto T (1992) Rabbit very low density lipoprotein receptor a low density lipoprotein receptor-like protein with distinct ligand specificity Proc Natl Acad Sci USA 89. 9252–9256

Terry RD, Masliah E, Salmon DP (1991) Physical basis of cognitive alterations in Alzheimer's disease· synapse loss is the major correlate of cognitive impairment Ann Neurol 41: 572–580

Tooyama I, Kawamata T, Akiyama H, Moestrup SK, Gliemann J, McGeer PL (1993) Immunohistochemical study of α2 macroglobulin receptor in Alzheimer and control postmortem human brain. Mol Chem Neuropathol 18· 153–160

Tumilowicz JJ, Nichols WW, Cholon JJ, Greene AE (1970) Definition of a continuous human cell line derived from neuroblastoma. Cancer Res 30· 2110–2118

Ueki A, Kawano M, Namba Y, Kawakami M, Ikeda K (1993) A high frequency of apolipoprotein E4 isoprotein in Japanese patients with late-onset nonfamilial Alzheimer's disease Neurosci Lett 163 166–168

Webb JC, Patel DD, Jones MD, Knight BL, Soutar AK (1994) Characterization and tissue-specific expression of the human "very low density lipoprotein (VLDL) receptor" mRNA Human Mol Genet 3 531–537

Wisniewski T, Frangione B (1992) Apolipoprotein E a pathological chaperone protein in patients with cerebral and systemic amyloid. Neurosci Lett 135 235–238

Wisniewski T, Golabek A, Matsubra E, Frangione B (1993) Apolipoprotein E binding to soluble Alzheimer's β-amyloid Biochem Biophys Res Commun 192 359–365

Wisniewski T, Catano EM, Golabek A, Vogel T, Frangione B (1994) Accerelation of Alzheimer's fibril formation by apolipoprotein E in vitro Am J Pathol 145 1030–1035

Wolf BB, Lopes MBS, VandenBerg SR, Gonias SL (1992) Characterization and immunohistochemical localization of α2-macroglobulin receptor (low-density lipoprotein receptor-related protein) in human brain Am J Pathol 141. 37–42

Yamamoto T, Takahashi S, Sakai J, Kawarabayasi Y (1993) The very low density lipoprotein receptor a second lipoprotein receptor that may mediate uptake of fatty acids into muscle and fat cells. TCM 3. 144–148

Yamazaki T, Selkoe DJ, Koo EH (1995) Trafficking of cell surface β-amyloid precursor protein retrograde and transcytotic transport in cultured neurons J Cell Biol 129 431–442

Yang B, Yang BL, Savani RC, Turley EA (1994) Identification of a common hyaluronate binding motif in the hyaluronate binding proteins RHAMM, CD44 and link protein EMBO J 13 286–296

Young ID, Willmer JP, Kisilevsky R (1989) The ultrastructural localization of sulfated proteoglycans is identical in the amyloids of Alzheimer's disease and AA, AL, senile cardiac and medullary carcinoma-associated amyloidosis Acta Neuropathol 78 202–209

Apolipoprotein E and Apolipoprotein E Receptors in the Central Nervous System: Relationship with Alzheimer's Disease

B.T. Hyman, G.W. Rebeck, R. Christie, K. Page, M. Irizarry,* and *D.K. Strickland*

Summary

With the recent discovery that apolipoprotein E (apoE) ε4 is in genetic disequilibrium with Alzheimer's disease and appears to be a risk factor for developing Alzheimer's disease, there has been intense interest in understanding the biological mechanisms underlying this genetic predisposition. Apolipoprotein E has been extensively studied in the context of its role in lipid metabolism in plasma, where it binds to lipoproteins and directs them to several specific receptor molecules. These receptor molecules internalize apoE and its bound ligands, frequently for degradation. We have begun to explore the expression of apoE and its receptors in the central nervous system as a first approximation to understanding the role of apoE in the pathophysiology of Alzheimer's disease. We have found that three different apoE receptors are expressed in the central nervous system. These receptors are expressed in different cell types and appear to be differentially regulated. Two of them, the low density lipoprotein receptor related protein (LRP) and the very low density lipoprotein receptor (VLDL-r) are located in the vicinity of apoE/Aβ deposit in the Alzheimer's brain. This anatomical relationship suggests hypotheses regarding the interactions of apoE, Aβ, and apoE receptors.

Introduction

The apolipoprotein E (apoE) protein has 299 amino acids and two major domains: a ligand binding domain that consists of amphipathic alpha helices that interact with a variety of amphipathic and hydrophobic substances, and a receptor binding domain that interacts with members of the LDL receptor family of receptor proteins. In the periphery, apoE appears to act as a transport protein mediating lipid and cholesterol transport between the liver and other organs.

A role for apoE in Alzheimer's disease began to emerge with the observation that apoE immunoreactivity occurs in cerebral amyloid deposits and

* Neurology Service, Massachusetts General Hospital, Boston, MA 02114, USA

A D Roses et al (Eds)
Apolipoprotein E and Alzheimer's Disease

neurofibrillary tangles (Namba et al. 1991). Wisniewski and Frangione (1992) suggested that apoE could act as a pathological chaperone for amyloid deposition. It had also been observed that astrocytic apoE mRNA was increased after lesion in the central nervous system (Poirier et al. 1991) and in Alzheimer's disease (Diedrich et al. 1991). The major spark of interest, however, came from the observation that the apoE ε4 allele was over-represented in Alzheimer's disease (Payami et al. 1993; Poirier et al. 1993; Rebeck et al. 1993; Strittmatter et al. 1993). The role of apoE in the central nervous system had been postulated to be a way of clearing myelin and other hydrophobic substances following nerve lesion (Mahley 1988). More recent data suggest that apoE supports neurite outgrowth when complexed to lipid particles, and that the apoE3 isoform supports neurite outgrowth to a greater extent than the apoE4 isoform (Nathan et al. 1994).

Clinical–Pathological Correlates of ApoE Genotypes in Alzheimer's Disease

It is now clear that inheritance of the ε4 allele of the apoE gene is a risk factor for Alzheimer's disease (Poirier et al. 1993; Rebeck et al. 1993; Strittmatter et al. 1993; Table 1). We have examined the clinical and pathological correlates of inheritance of various apoE alleles and have found that there are selective effects on specific aspects of the disease process. For example, individuals who inherit apoE ε4 are more likely to have an earlier age of onset (Poirier et al. 1993; Corder et al. 1994; Tsai et al. 1994; West et al. 1994). However, some individuals are able to live to extreme old age and remain cognitively normal into their 90's, despite inheritance of an apoE ε4 allele (Rebeck et al. 1994). In recent studies, we have been unable to detect any change in the rate of progression of the disease process, as measured either by cognitive scores or activity of daily living scales (West et al. 1994), as a function of apoE genotype.

From a neuropathological prospective, we and others have demonstrated that there is increased amyloid deposition in the brains of individuals who have inherited apoE ε4 (Rebeck et al. 1993; Berr et al. 1994; Schmechel et al.

Table 1. ApoE allele frequencies in Alzheimer's disease[a]

ApoE	Control (n = 129)	Alzheimer's disease (n = 350)
ε2	0 08	0 04**
ε3	0 78	0 57
ε4	0 14	0 39**

[a] Genotypes derived from a study of Alzheimer disease or control individuals from the Massachusetts Alzheimer Disease Research Center

$^{**}p < 0.01$, chi-squared goodness of fit test

1993). Increased amyloid deposition in terms of Congophilic angiopathy is particularly increased in apoE $\varepsilon 4$ individuals. We find no effect of apoE genotype on neurofibrillary tangle number or rate of accumulation (West et al. 1994).

From the observation that there was increased amyloid deposition associated with inheritance of apoE $\varepsilon 4$, we suggested that apoE acts as a clearance mechanism for the small, amphipathic peptide Aβ and we proposed that ApoE4 was somehow less efficient at clearing Aβ than apoE3 (Rebeck et al. 1993). The idea of a clearance mechanism implied that apoE receptors were present in the central nervous system and were in a position to interact with apoE/Aβ complexes. Earlier work had demonstrated the presence of the LDL receptor in the central nervous system (Pitas et al. 1987). However, in the last several years a family of LDL receptor-like proteins has been discovered including the LDL receptor related protein (LRP; Strickland et al. 1990), the very low density lipoprotein receptor (VLDLr; Takahashi et al. 1992), and Megalin (GP330; Saito et al. 1994).

Expression of ApoE Receptors in the Central Nervous System

We have begun to explore the expression of these receptors in the adult central nervous system in both rats and humans (Table 2). Because all of the receptors have been cloned and sequenced, both in situ hybridization and immunohistochemical techniques are available. By Western blot assay, we found abundant LDL receptor and LRP in human cerebral cortex. VLDL-r is also present, although in apparently lower amounts. We have been unable to detect GP330. In situ hybridization studies in the adult rat show parallel results.

Immunohistochemical analysis reveals diffuse LDL receptor staining with both monoclonal and polyclonal antibodies, suggesting a diffuse and perhaps universal role for this receptor. LRP, by contrast, is present primarily on large neurons, especially within the hippocampal formation. In addition, we find

Table 2. ApoE receptors in the central nervous system

Name	Abbreviation	Location
Low-density lipoprotein receptor	LDL-r	Abundant, diffuse neuropil staining
Low-density lipoprotein receptor protein; α_2-macroglobulin receptor	LRP	Abundant, large pyramidal neurons; reactive astrocytes, senile plaques
Very low density lipoprotein receptor	VLDL-r	Microglia, occasional pyramidal neuron
Glycoprotein 330; Megalin	GP330	Absent in human cortex by immunostaining; ? choroid plexus, ependyma

LRP on reactive, but not on resting, astrocytes. Finally, in Alzheimer's disease, we find LRP in association with senile plaques (Rebeck et al. 1993). VLDL receptor is present on a small population of large pyramidal neurons, as well as in a striking fashion on reactive microglial cells. In Alzheimer's disease, these reactive microglia surround senile plaques. Thus these various receptors are expressed in different cell populations and appear also to be regulated in various ways. Both LRP and VLDL receptor are associated with senile plaques in Alzheimer's disease.

LRP is of particular interest because it appears to be directly related to the majority of cortical Aβ deposits. LRP is a multifunctional receptor, with binding sites for at least seven known ligands. These ligands appear to act independently of one another and include apoE, lactoferrin, PAI-1, tPA, uPA, alpha 2 macroglobulin, and lipoprotein lipase. Knowing that both lactoferrin and alpha 2 macroglobulin had previously been described as plaque-associated proteins, we wondered whether other LRP ligands might also be associated with plaques. Using immunohistochemical techniques we recently demonstrated that all known LRP ligands, a diverse group of otherwise apparently unrelated molecules, are present on at least some senile plaques (Rebeck et al. 1995).

Concluding Comments

Much remains to be learned about the physiology and metabolism of apoE in the central nervous system. We have discovered that several different members of the LDL receptor family of apoE receptors are expressed in the adult brain, although they are expressed on different cell types and apparently under different circumstances. There is little doubt that these receptors are regulated in distinct ways. It remains possible that there are isoform-specific differences in apoE's interactions with receptors that could contribute to the biological underpinnings of the apoE allele effect seen in genetic studies. Our ongoing studies aim to further define the role that various receptors play in the central nervous system and under normal conditions after injury. It is our hope that by understanding apoE and apoE receptor physiology and pharmacology better, we will be able to develop strategies to intervene in the early pathophysiological processes of Alzheimer's disease.

Acknowledgments. We thank Sharon Melanson for preparing the manuscript. Supported by NIH grants AG12406-01 and AG05134-11.

References

Berr C, Hauw J-J, Delaere P, Duyckaerts C, Amouyel P (1994) Apolipoprotein E allele ε4 is linked to increased deposition of the amyloid β-peptide (A-β) in cases with or without Alzheimer's disease. Neurosci Lett 178· 221–224

Corder EH, Saunders AM, Risch NJ, Strittmatter WJ, Schmechel DE, Gaskell PC, Rimmler JB, Locke PA, Conneally M, Schmader KE, Small GW, Roses AD, Haines JL, Pericak-Vance MA (1994) Apoliporotein E type 2 allele decreases the risk of late onset Alzheimer disease Nature Genet 7 180–184

Diedrich JF, Minnigan H, Carp RI, Whitaker JN, Race R, Frey W III, Haase AT (1991) Neuropathological changes in scrapie and Alzheimer's disease are associated with increased expression of apoliporotein E and cathepsin D in astrocytes J Virol 65 4759–4768

Mahley RW (1988) Apolipoprotein E cholesterol transport protein with expanding role in cell biology Science 240 622–630

Namba Y, Tomonaga M, Kawasaki H, Otomo E, Ikeda K (1991) Apolipoprotein E immunoreactivity in cerebral amyloid deposits and neurofibrillary tangles in Alzheimer's disease and kuru plaque amyloid in Creutzfeldt-Jakob disease Brain Res 541: 163–166

Nathan BP, Bellosta S, Sanan DA, Weisgraber KH, Mahley RW, Pitas RE (1994) Differential effects of apolipoproteins E3 and E4 on neuronal growth in vitro Science 264 850–852

Payami H, Kaye J, Heston LL, Bird TD, Schellenberg GD (1993) Apolipoprotein E genotype and Alzheimer's disease (letter) Lancet 342 738

Pitas RE, Boyles JK, Lee SH, Foss D, Mahley RW (1987) Astrocytes synthesize apolipoprotein E and metabolize apolipoprotein E-containing lipoproteins Biochim Biophys Acta 917 148–161

Poirier J, Hess M, May PC, Finch CE (1991) Astrocytic apolipoprotein E mRNA and GFAP mRNA in hippocampus after entorhinal cortex lesioning. Mol Brain Res 11 97–106

Poirier J, Davignon J, Bouthillier D, Kogan S, Bertrand P, Gauthier S (1993) Apolipoprotein E polymorphism and Alzheimer's disease Lancet 342. 697–699

Rebeck GW, Reiter JS, Strickland DK, Hyman BT (1993) Apolipoprotein E in sporadic Alzheimer's disease: allelic variation and receptor interactions Neuron 11 575–580

Rebeck GW, Perls TT, West HL, Sodhi P, Lipsitz LA, Hyman BT (1994) Reduced apolipoprotein E4 allele frequency in the oldest old Alzheimer's patients and cognitively normal individuals Neurology 44 1513–1516

Rebeck GW, Harr SD, Strickland DK, Hyman BT (1995) Multiple, diverse senile plaque-associated proteins are ligands of an apolipoprotein E receptor, the α2-macroglobulin receptor/low-density-lipoprotein receptor-related protein Ann Neurol 37· 211–217

Saito A, Pietromonaco S, Loo AK-C, Farquhar MG (1994) Complete cloning and sequencing of rat gp330/"megalin," a distinctive member of the low density lipoprotein receptor gene family Proc Natl Acad Sci USA 91: 9725–9729

Schmechel DE, Saunders AM, Strittmatter WJ, Crain B, Hulette C, Joo SH, Pericak-Vance MA, Goldgaber D, Roses AD (1993) Increased amyloid β-peptide deposition as a consequence of apolipoprotein E genotype in late-onset Alzheimers disease Proc Natl Acad Sci 90 9649–9653

Strickland DK, Ashcom JD, Williams S, Burgess WH, Migliorini M, Argraves WS (1990) Sequence identity between the α2-macroglobulin receptor and low density lipoprotein receptor-related protein suggests that this molecule is a multifunctional receptor. J Biol Chem 265· 17401–17404

Strittmatter WJ, Saunders AM, Schmechel D, Pericak-Vance M, Enghild J, Salvesen GS, Roses AD (1993) Apolipoprotein E: high-avidity binding to β-amyloid and increased frequence of type 4 allele in late-onset familial Alzheimer disease. Proc Natl Acad Sci USA 90· 1977–1981

Takahashi S, Kawarabayasi Y, Nakai T, Sakai J, Yamamoto T (1992) Rabbit very low density lipoprotein receptor. a low density lipoprotein receptor-like protein with distinct ligand specificity Proc Natl Acad Sci USA 89 9252–9256

Tsai M-S, Tangalos EG, Petersen RC, Smith GE, Schaid DJ, Kokmen E, Ivnek RJ, Thibodeau N (1994) Apoliprotein E risk factor for Alzheimer's disease Am J Human Genet 54 643–649

West H, Rebeck GW, Growdon JH, Hyman BT (1994) Apolipoprotein E affects neuropathology but not clinical progression in Alzheimer's disease Proc Fourth Intl Conference on Alzheimer's Disease, in press

Wisniewski T, Frangione B (1992) Apolipoprotein E a pathological chaperone protein in patients with cerebral and systemic amyloid Neurosci Lett 135 235–238

Tau Protein and the Neurofibrillary Pathology of Alzheimer's Disease

*M. Goedert**

Introduction

Alzheimer's disease is characterised clinically by a progressive loss of memory and other cognitive functions, resulting in a profound dementia. The intellectual decline is accompanied by the progressive accumulation in the brain, both extracellularly and intracellularly, of insoluble fibrous material in the form of senile plaques and neurofibrillary lesions. A diagnosis of Alzheimer's disease is made when an elderly patient exhibits clinical evidence of a chronic progressive dementia and when post-mortem examination of the brain reveals abundant neuritic plaques and neurofibrillary lesions in cerebral cortex and hippocampal formation. Although the existence of these abnormal deposits has been known since the first description of the disease by Alzheimer in 1907, only in the past 12 years have they yielded to an intensive molecular analysis (for reviews, see Goedert 1993; Selkoe 1994). Parallel lines of investigation have concentrated on familial forms of the disease, with the aim of discovering the genes whose malfunction causes the disease in these families. Although the picture is still incomplete, it is already clear that Alzheimer's disease is a genetically heterogeneous disorder, with different primary causes leading to the same phenotype and neuropathology (Fig. 1; for reviews, see Mullan and Crawford 1993; Strittmatter et al. 1994a). It is therefore likely that the activation of several distinct pathological pathways can lead to the disease, with neuritic plaques and neurofibrillary lesions representing the points of convergence of these events. The formation of the latter is believed to produce the symptoms of the disease, which result most probably from the degeneration of nerve cells in cerebral cortex and hippocampal formation, with ensuing neuronal loss and reduction in synapse numbers (Fig. 1). Therefore, a study of the mechanisms that lead to plaque and tangle formation is essential for an understanding of the pathogenesis of all forms of Alzheimer's disease.

Extracellular deposits are composed of amyloid fibrils whose major component is the β-amyloid protein Aβ (Glenner and Wong 1984; Masters et al. 1985). Initial deposits are diffuse, whereas more mature deposits give rise to the characteristic senile plaques. Neurofibrillary lesions constitute the

* MRC Laboratory of Molecular Biology, Hills Road, Cambridge CB2 2QH, UK

A D Roses et al (Eds)
Apolipoprotein E and Alzheimer's Disease

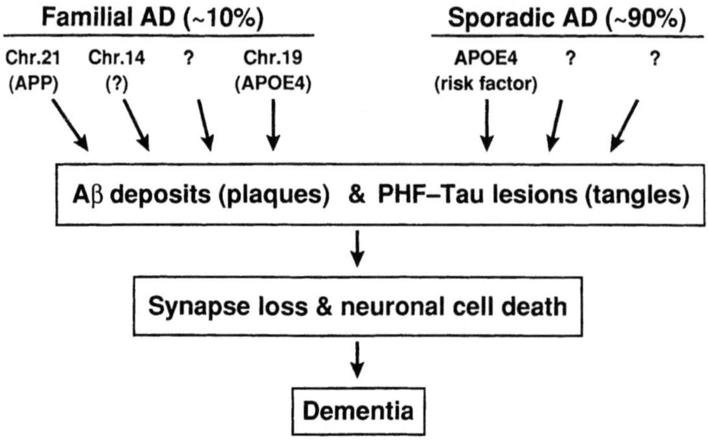

Fig. 1. Aetiological heterogeneity of Alzheimer's disease (*AD*) Summary of known aetiologies and risk factors for familial and sporadic AD Three chromosomal loci, chromosomes (*Chr*) 21, 14 and 19, have been identified, point mutations in the amyloid precursor protein (*APP*) are associated with a small number of early-onset familial cases. The majority of early-onset familial cases are caused by an as yet unidentified gene defect on chromosome 14 Apolipoprotein E genotype is a risk factor for some familial cases of AD and for the sporadic forms of the disease Pathogenic mechanisms converge, resulting in the formation of β-amyloid protein ($A\beta$) deposits and neurofibrillary lesions This process in turn is believed to lead to the neuronal loss characteristic of AD *APOE*, apolipoprotein E, *PHF*, paired helical filament

intraneuronal deposits. They are found in nerve cell bodies and apical dendrites as neurofibrillary tangles, in distal dendrites as neuropil threads and in the abnormal neurites that are associated with some amyloid plaques (neuritic plaques; Fig. 2). Ultrastructurally, all three lesions contain abnormal paired helical filaments (PHFs) as their major fibrous components and straight filaments (SFs) as their minor fibrous components (Fig. 3; Kidd 1963). Neurofibrillary lesions develop in the vast majority of nerve cells that undergo degeneration in Alzheimer's disease. Their relative insolubility enables them to survive after the death of the affected nerve cells as extracellular tangles (or ghost tangles) that accumulate in the neuropil. These are then engulfed by astrocytes and microglial cells and are probably slowly degraded.

Recently, significant progress has been made in unravelling the molecular composition of PHFs and in deducing possible mechanisms that may lead to their assembly. Current evidence strongly suggests that they are made entirely of microtubule-associated protein tau in a hyperphosphorylated state. Moreover, earlier results indicating that the extent and topographical distribution of neurofibrillary lesions provide a reliable pathological correlate of the degree of dementia have been confirmed and extended (Braak and Braak 1991; Arriagada et al. 1992). The finding of large numbers of amyloid deposits in some cognitively normal individuals (also known as "pathological ageing") has shown that amyloid deposits are not sufficient for dementia (Fig. 4) (Dickson

Fig. 2. Neurofibrillary pathology in the entorhinal cortex The section was stained with an anti-tau antiserum. *NFT*, neurofibrillary tangle: *NT*, neuropil threads, *NP*, neuritic plaque *Scale bar*, 80 μm

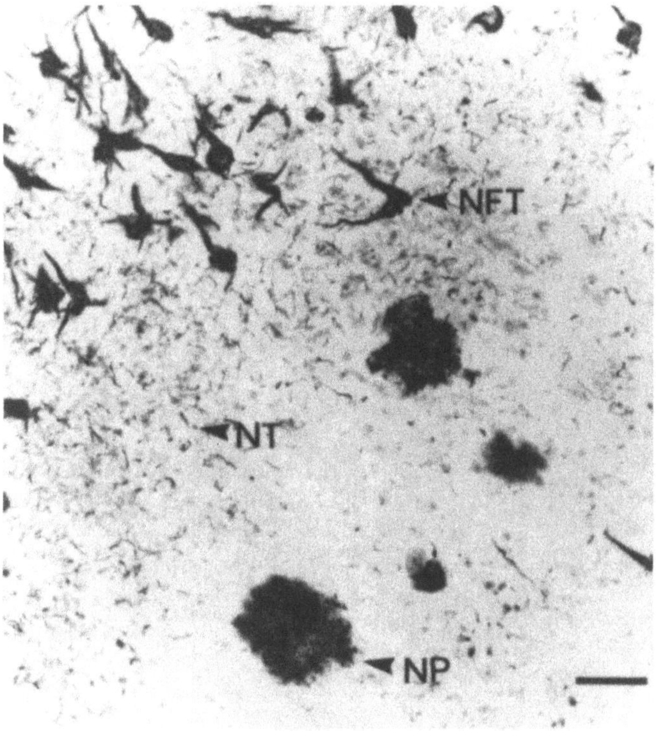

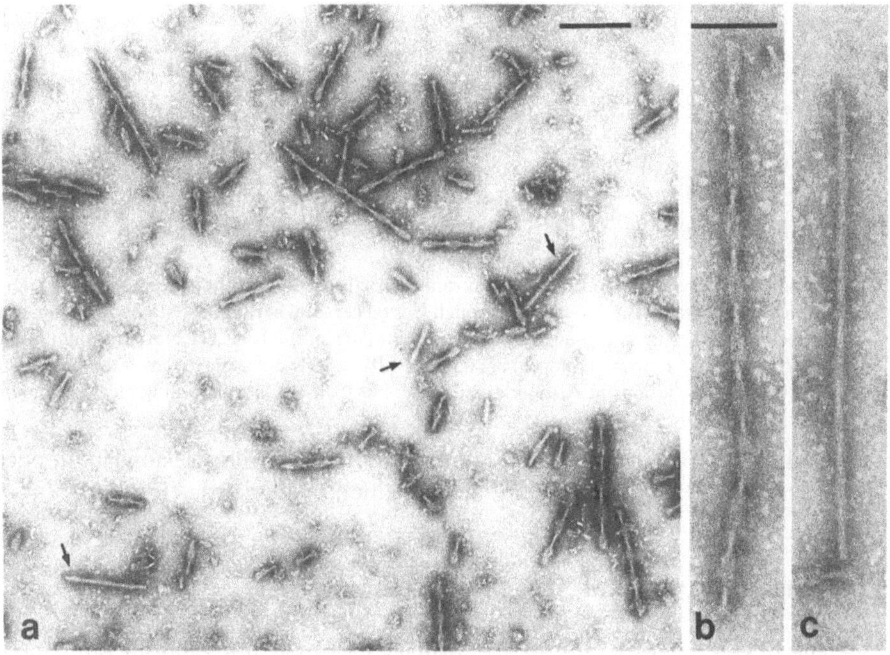

Fig. 3. Electron micrographs of negatively stained abnormal filaments from the brain of an Alzheimer's disease brain **a** low-power view showing predominantly paired helical filaments but with a few straight filaments (*arrows*) **b,c** high-power views of a paired helical filament (**b**) and a straight filament (**c**) *Scale bars*, 200 nm (**a**), 75 nm (**b,c**)

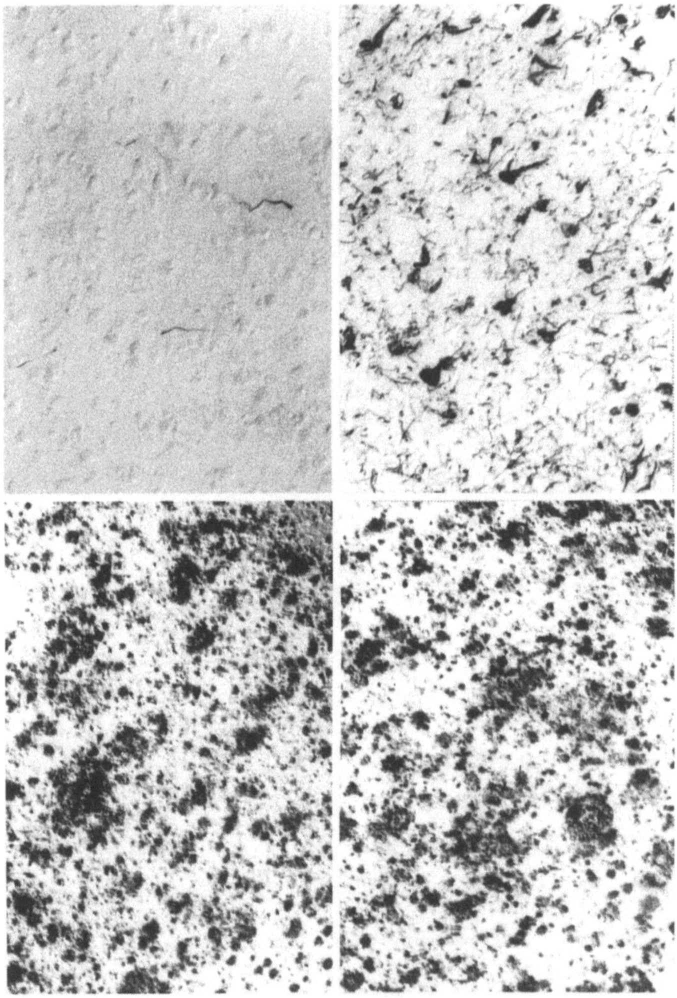

Fig. 4A–D. Amyloid deposits and neurofibrillary lesions in pathological ageing (**A,C**) and Alzheimer's disease (**B,D**) Sections through frontal cerebral cortex were stained with an anti-tau antibody (**A,B**) or an anti-β-amyloid protein (anti-Aβ) antibody (**C,D**). Note the similar density of Aβ deposits in both pathological ageing and Alzheimer's disease Note also the large number of neurofibrillary lesions in Alzheimer's disease, but their absence in pathological ageing

et al. 1991). By contrast, large numbers of neurofibrillary lesions in hippocampus and/or cerebral cortex are accompanied by dementia (Fig. 4).

Neuropathological Stages of Alzheimer's Disease

The development of the neurofibrillary lesions is not random but follows a stereotyped pattern with regard to affected cell types, cellular layers and brain regions, with little individual variation. This finding has been used to define six neuropathological stages of Alzheimer's disease (Braak and Braak 1991). The very first nerve cells in the brain to develop neurofibrillary lesions are located in layer pre-alpha of the transentorhinal region, thus defining stage I. Stage II shows a more severe involvement of this region, as well as a mild involvement of the pre-alpha layer of the entorhinal cortex. Patients with this pathology are cognitively unimpaired, indicating that stages I and II may represent clinically silent stages of Alzheimer's disease. Mild impairments of cognitive function become apparent in stages III and IV. Stage III is characterised by severe neurofibrillary lesions in the pre-alpha layers of both entorhinal and transentorhinal regions. The vast majority of nerve cells shows neurofibrillary tangles and dendritic neuropil threads. The first extracellular tangles also appear during stage III. In stage IV the deep pre-alpha layer develops extensive neurofibrillary lesions. During stages III and IV mild changes are also seen in layer I of Ammon's horn of the hippocampus and in a number of subcortical nuclei, such as the basal forebrain magnocellular nuclei and the anterodorsal thalamic nucleus. The major feature of stages V and VI is the massive development of neurofibrillary lesions in isocortical association areas. They meet the criteria for the neuropathological diagnosis of Alzheimer's disease and are found in patients who were severely demented at the time of death.

The stereotyped nature of the temporal and spatial development of neurofibrillary lesions contrasts with the development of Aβ deposits. They show a density and distribution pattern that are subject to great individual variation, precluding their use for the neuropathological staging of Alzheimer's disease. In general, the first Aβ deposits occur in isocortical areas of the frontal, temporal and occipital lobes. This contrasts with the neurofibrillary lesions, which first appear in the transentorhinal region. Moreover, Aβ deposits develop relatively late in the facia dentata of the hippocampus, the major termination area of the pre-alpha layer cells of the entorhinal cortex. It follows that neurofibrillary lesions in the pre-alpha layer can develop with no parts of these cells or their processes in contact with Aβ deposits. These findings are inconsistent with the view that the neurofibrillary pathology develops as a mere consequence of the neurotoxic action of Aβ (Kowall et al. 1991).

Structure of the Paired Helical Filaments

The PHF, as its name suggests, consists of two strands of subunits that twist around one another in a helical fashion (Fig. 3). When viewed in the electron microscope, the helical twist and relative disposition of the two strands give rise to images in which the width alternates between about 8 and 20 nm, with an apparent period of 80 nm (Crowther and Wischik 1985).

PHFs can be isolated either in the form of tangle fragments (Wischik et al. 1988; Kondo et al. 1988) or dispersed filaments (Greenberg and Davies 1990;

Lee et al. 1991). The two types of PHFs have tau epitopes in common but differ in their solubility in strong denaturing agents. Although a majority of dispersed PHFs are soluble in guanidine or sodium dodecylsulphate, a majority of tangle fragment PHFs are insoluble in these reagents. Both types of PHFs differ also in their sensitivity to proteases. Pronase treatment of tangle preparations removes a fuzzy coat from the PHF and leaves behind a pronase-resistant core. The morphology of the core is similar to that of untreated PHFs, but structural details are seen more clearly because the disordered coat has been removed. From electron micrographs it is possible to compute a map of the cross-sectional density in the core. Such maps show two C-shaped morphological units, corresponding to the two strands of the PHF, arranged in a base-to-base manner (Crowther and Wischik 1985). When dispersed filaments are treated with pronase under the same conditions, they are completely degraded (Goedert et al. 1992a), illustrating the differing protease sensitivities of tangle fragment PHFs and of dispersed PHFs.

Straight filaments represent a minority species, both in tangle fragment and dispersed filament preparations. Images of SFs show approximately the same apparent periodicity as PHFs, but a much less marked modulation in width (Fig. 3; Crowther 1991). Straight filaments and PHFs share tau epitopes and behave in a similar manner when treated with pronase. Rarely, hybrid filaments are observed that show a sharp transition from a segment of PHF into a segment of SF. This indicates that PHFs and SFs contain identical or closely related subunits that are arranged differently in the two types of filament. This is supported by the computed cross-section of the SF, which shows two C-shaped subunits very similar to those seen in the PHF, but arranged back-to-back rather than base-to-base (Crowther 1991). The SF is thus a structural variant of the PHF, in that both contain two strands of closely related or possibly identical subunits but the relative arrangement of the two strands differs in the two kinds of filament.

Tau Protein

Tau is an abundant protein in both central and peripheral nervous systems (for review, see Goedert et al. 1994a). In brain it is found predominantly in nerve cells, where it is concentrated in nerve cell axons (Binder et al. 1986; Peng et al. 1986). This contrasts with the distribution of the neurofibrillary lesions in Alzheimer's disease, which are found in nerve cell bodies, dendrites and axons. Molecular cloning and sequencing of tau cDNAs (Lee et al. 1988; Goedert et al. 1988, 1989a) have shown that multiple tau isoforms are produced from a single gene through alternative mRNA splicing. In adult human brain six isoforms are found that range from 352 to 441 amino acids and differ from each other by the presence or absence of three inserts (Fig. 5; Goedert et al. 1989a; Goedert and Jakes 1990). The most striking feature of the tau sequences is the presence of three or four tandem repeats of 31 or 32 amino acids located in the carboxy-terminal half (Goedert et al. 1989b). Experiments with recombinant tau proteins show that the repeats constitute microtubule-binding domains (Butner and Kirschner 1991; Goode and Feinstein 1994). Microtubules assembled in the presence of tau show arms projecting from the surface (Hirokawa et al. 1988). Tau thus consists of a carboxy-terminal microtubule-binding domain and an amino-terminal projection domain. Besides being distinguished by the presence of three or four tandem repeats, some tau isoforms contain 29 or 58 amino acid inserts located near the amino-terminus (Fig. 5). Isoforms with a large additional insert in the amino-terminal half have been described in the peripheral nervous system (Goedert et al. 1992b; Couchie et al. 1992).

Tau is subject to developmentally regulated alternative mRNA splicing in that, in immature brain, only the transcript encoding the shortest isoform with three repeats is expressed (Goedert et al. 1989a). The developmental shift of tau bands from a simple fetal pattern to a more complex adult pattern thus involves the transition from the expression of the isoform with three repeats and no inserts to the expression of all six isoforms.

Tau is a phosphoprotein (Butler and Shelanski 1986) and phosphorylation is also developmentally regulated. Thus, tau from immature brain is phosphorylated at more sites than tau from adult brain, implying selective dephosphorylation of the shortest isoform during brain maturation (Fig. 6). Tau from newborn brain is phosphorylated at 12 known sites in the shortest isoform, whereas in adult brain it is phosphorylated at at least six sites in all six isoforms. Phosphorylation sites have been identified through the use of mass spectrometry (Poulter et al. 1993; Watanabe et al. 1993) and of phosphorylation-dependent anti-tau antibodies (Figs. 7, 8; Kanemaru et al. 1992; Goedert

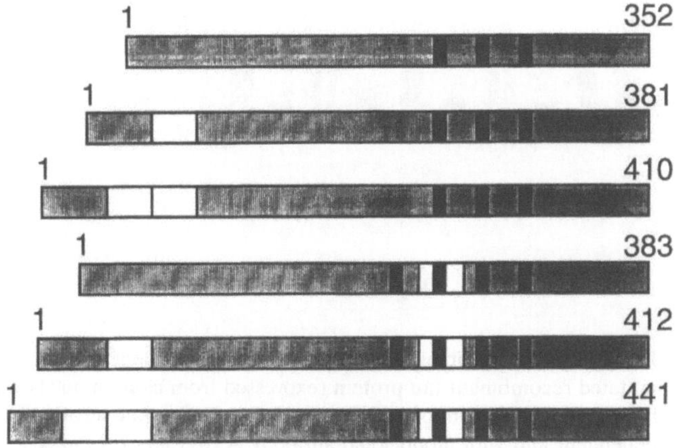

Fig. 5. The six human brain tau isoforms The region common to all isoforms is *stippled* and the inserts that distinguish them are shown in *white* The three or four tandem repeats are shown by *black bars* The number of amino acids is indicated The 352 amino acid isoform is expressed in foetal brain, whereas all six isoforms are expressed in adult brain

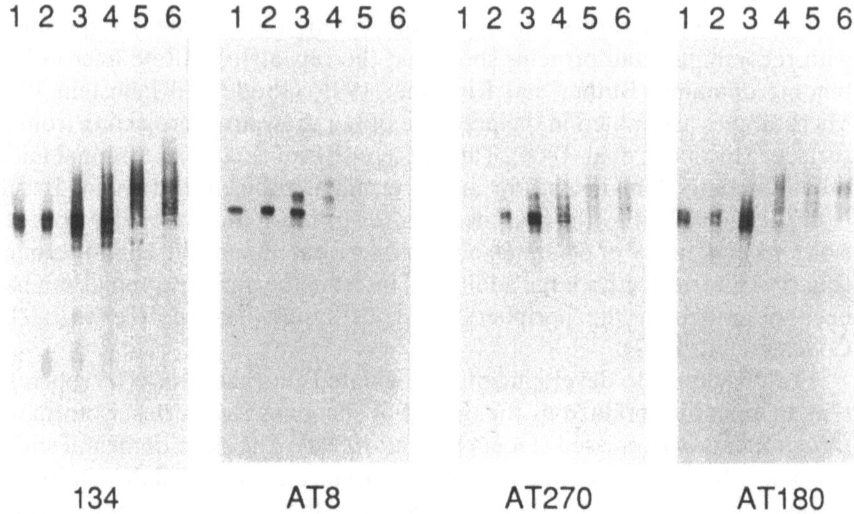

Fig. 6. Immunoblots of developing and adult rat brain with anti-tau serum 134 and phosphorylation-dependent monoclonal anti-tau antibodies AT8, AT270 and AT180 Serum 134 recognises the carboxy-terminus of tau and is not phosphorylation dependent. AT8 recognises tau protein phosphorylated at S202 and T205, AT270 recognises tau protein phosphorylated at T181, whereas AT180 recognises tau protein phosphorylated at T231 and S235 *Lane 1*, 19 days of gestation, *2*, 1 week old, *3*, 2 weeks old, *4*, 4 weeks old, *5*, 8 weeks old, *6*, 4 months old

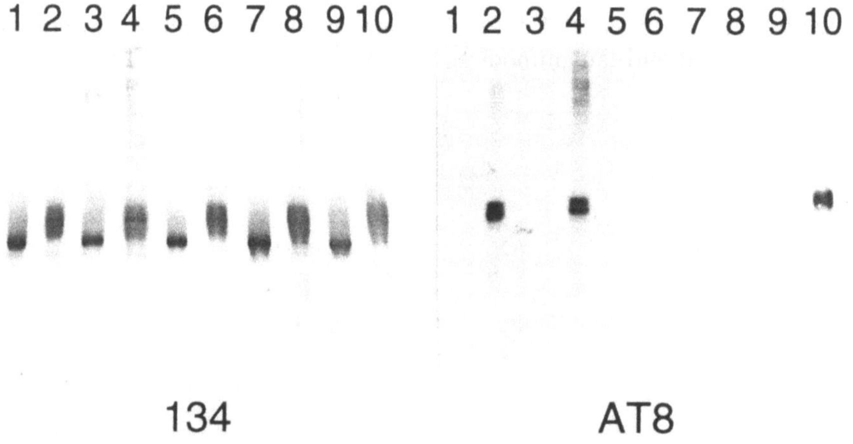

Fig. 7. Epitope mapping of phosphorylation-dependent anti-tau antibody AT8 Wild-type and mutated recombinant tau protein (expressed from clone htau24) was phosphorylated with a rat brain extract Immunoblots were stained with anti-tau serum 134 that is not phosphorylation dependent and with monoclonal antibody AT8 *Lane 1*, htau24; *2*, htau24 plus brain extract; *3*, S199A htau24, *4*, S199A htau24 plus brain extract, *5*, S202A htau24, *6*, S202A htau 24 plus brain extract, *7*, T205A htau24, *8*, T205A htau24 plus brain extract, *9*, T212A htau24, *10*, T212A htau24 plus brain extract These results show that AT8 specifically recognises tau protein phosphorylated at both S202 and T205

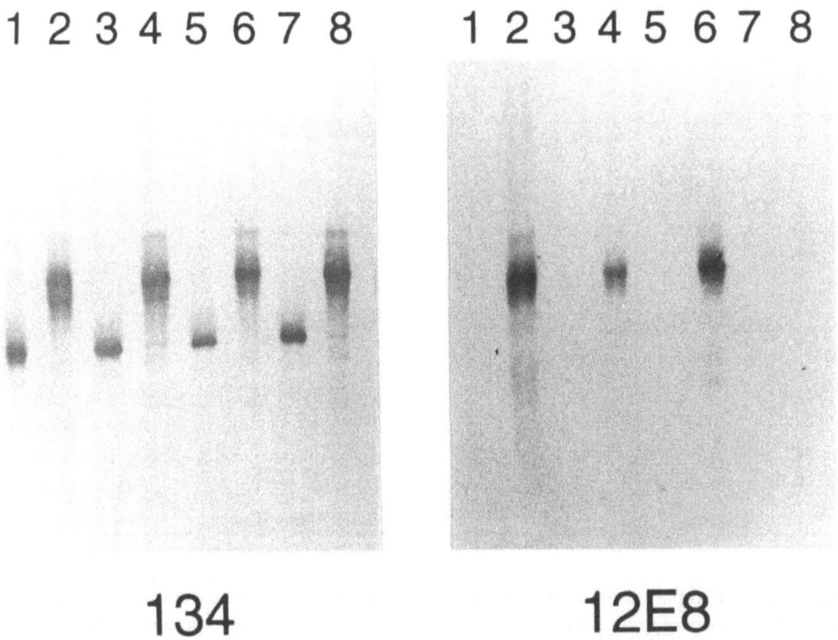

Fig. 8. Epitope mapping of phosphorylation-dependent anti-tau antibody 12E8 Wild-type and mutated recombinant tau protein (expressed from clone htau24) was phosphorylated with a rat brain extract Immunoblots were stained with anti-tau serum 134 that is not phosphorylation dependent and with monoclonal antibody 12E8 *Lane 1*, htau24, *2*, htau24 plus brain extract, *3*, S262A htau24, *4*, S262A htau24 plus brain extract, *5*, S356A htau24, *6*, S356A htau24 plus brain extract, *7*, S262A + S356A htau24, *8*, S262A + S356A htau24 plus brain extract These results show that 12E8 specifically recognises tau protein phosphorylated at S262 and/or S356

et al. 1993, 1994b, 1995; Bramblett et al. 1993; Kenessey and Yen 1993; Brion et al. 1993; Hasegawa et al. 1993; Matsuo et al. 1994; Seubert et al. 1995). Thus, in fetal brain, tau is phosphorylated at serine residues 198, 199, 202, 235, 262, 396, 400 and 404 and at threonine residues 181, 205, 217 and 231 (according to the numbering of the longest human brain tau isoform; Fig. 9). In adult brain tau is phosphorylated at serine residues 199, 202, 235, 262, 396 and 404 and at threonine residues 181, 205 and 231 (Fig. 9). With the exception of Ser262, all the known phosphorylation sites in tau are located outside the microtubule-binding repeat region. Phosphorylation is heterogeneous, implying that a given tau molecule is phosphorylated at some, but not all, of these sites. Many of these sites are serine and threonine residues that are followed by a proline, suggesting that protein kinases with a specificity for seryl-proline and threonyl-proline phosphorylate tau in normal brain. Accordingly, mitogen-activated protein (MAP) kinase, glycogen synthase kinase-3 (GSK3) and cyclin-dependent kinase 5 (cdk5) phosphorylate tau at a number of the above serine/threonine-proline residues in vitro (Drewes et al. 1992; Ledesma et al. 1992; Goedert et al. 1992c; Hanger et al. 1992; Mandelkow et al. 1992; Ishiguro

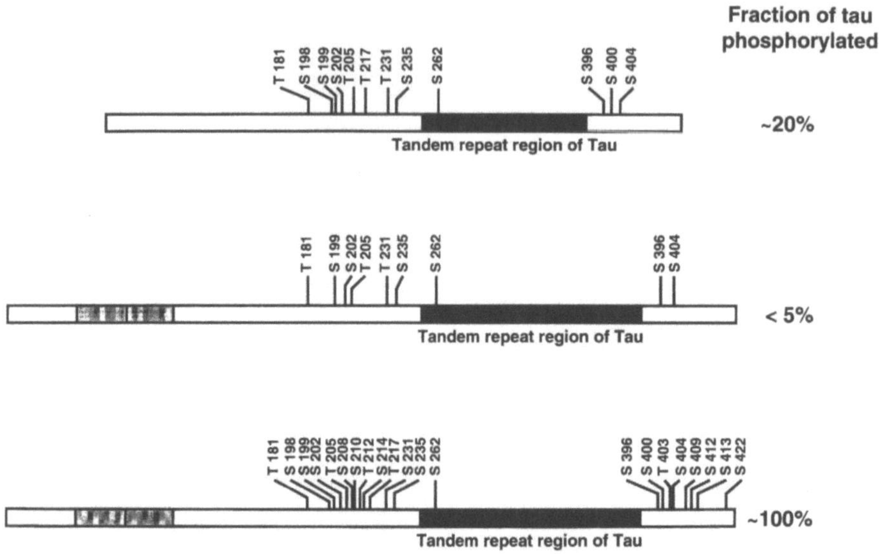

Fig. 9. Identified phosphorylation sites in tau from newborn brain (*top*), adult brain (*middle*) and Alzheimer's disease brain (*bottom*) In newborn brain only the shortest tau isoform is expressed, whereas six isoforms are expressed in adult human brain In adult brain and Alzheimer's disease brain, all six tau isoforms are phosphorylated on the residues shown here for the longest isoform The approximate fraction of tau that is phosphorylated at the sites shared between newborn brain, adult brain and Alzheimer's disease brain is shown at the *right*

et al. 1993; Paudel et al. 1993; Kobayashi et al. 1993; Baumann et al. 1993). In addition, cAMP-dependent protein kinase and Ca^{2+}/calmodulin-dependent protein kinase II phosphorylate serine residue 262 of tau in vitro (Litersky et al., submitted for publication), as does a partially purified 110kDa protein kinase from brain (Drewes et al. 1995).

The phosphorylation state of a protein is the result of a balance between protein kinase and protein phosphatase activities. Of the major phosphatase activities present in a crude brain extract tau phosphorylated by MAP kinase, GSK3 or cdk5 is dephosphorylated predominantly by the trimeric form of protein phosphatase 2A (Goedert et al. 1992c). The same has been shown for tau from adult human brain (Szücs et al. 1994). A substantial proportion of the cytosolic pool of protein phosphatase 2A from a number of neuronal and non-neuronal cell lines is associated with microtubules (Sontag et al. 1995).

Tau Functions

Ever since its discovery tau has been known to be a potent promoter of tubulin polymerisation in vitro (Weingarten et al. 1975). The binding of tau to microtubules reduces their dynamic instability (Drechsel et al. 1992). Analysis of the dynamics of the growth of individual microtubules indicates that tau increases the rate of association and decreases the rate of dissociation of tubulin molecules at the growing end and inhibits the transition to the catastrophic shortening phase. Binding studies have indicated that the affinity of the 31 or 32 amino acid repeat region for microtubules is concentrated in 18 amino acid-binding elements that are separated by flexible linker sequences (Butner and Kirschner 1991). However, recent results have shown that the picture is more complex, with at least one linker region contributing significantly towards microtubule binding (Goode and Feinstein 1994). This is in keeping with the highly conserved nature of the whole repeat region between species (for review, see Goedert et al. 1994a).

Tau has a similar function in vivo; when microinjected into fibroblasts it produces an increase in microtubule mass and an increased resistance of microtubules to depolymerising agents (Drubin and Kirschner 1986). Several laboratories have transfected various tau constructs into cells that do not normally express tau (Kanai et al. 1989, 1992; Knops et al. 1991; Lee et al. 1992; Gallo et al. 1992; Lo et al. 1993; Bramblett et al. 1993). In the transfected cells tau binds to microtubules, enhances microtubule stability and, in some cases, induces microtubule bundling. The bundle thickness is ten to 100 times that of normal microtubules, with an average distance of about 20nm between adjacent microtubules. Bundling is probably a consequence of the microtubule stabilisation effected by tau. At present, it is unclear whether microtubule bundling has a physiological correlate or whether it results from the non-physiological overexpression of tau in some cell types. Experiments with truncated tau proteins have confirmed the critical role played by the tandem repeat region in microtubule binding; however, the inclusion of sequences flanking the repeats appears to be essential for optimal binding and bundling. In particular, the proline-rich region upstream of the repeats appears to be essential for tubulin binding in vivo; this region contains a cluster of serine/threonine-proline phosphorylation sites.

In Sf9 cells infected with a baculovirus construct, the overexpression of tau results not only in microtubule bundles but also in the extension of long and thin neurite-like extensions (Knops et al. 1991). However, overexpression of tau does not invariably lead to the extension of neurite-like processes. This might be due to different levels of expression and phosphorylation in the various cell types. Tensile forces exerted by the cortical actin network are likely to represent another factor. Thus, cells transfected with the microtubule-associated protein MAP2c only grow processes when treated with the actin depolymerising drug cytochalasin B (Edson et al. 1993).

The introduction of antisense oligonucleotides into cultured developing cerebellar nerve cells to block tau expression prevents the differentiation of the short neurites, one of which would otherwise have developed into an axon (Caceres and Kosik 1990). Similarly, tau antisense oligonucleotides produce a retraction of neurites in PC12 cells treated with nerve growth factor (Hanemaaijer and Ginzburg 1991). By contrast, mice with an inactivating mutation in the tau gene show normal brain development (Harada et al. 1994);

moreover, cultured developing cerebellar nerve cells from these animals grow a normal complement of axons and dendrites, demonstrating that tau protein is not essential for process outgrowth and axonal morphology (Harada et al. 1994).

Phosphorylation negatively regulates the ability of tau to bind to microtubules. Tau phosphorylated by MAP kinase has one-tenth the ability of non-phosphorylated tau to bind to microtubules (Drechsel et al. 1992). Little is known about the relative contributions made by individual phosphorylated residues. A study using transfected cells has shown that phosphoserine residue 396 makes a significant contribution towards the reduced ability of tau to bind to microtubules (Bramblett et al. 1993). This and similar approaches should permit identification of the other sites involved.

Tau Protein and the Paired Helical Filament

Protein chemistry and molecular cloning established directly that tau forms an integral component of the PHF (Goedert et al. 1988; Wischik et al. 1988; Kondo et al. 1988), thereby putting more indirect immunohistochemical studies (Brion et al. 1985) on a solid basis. The tau sequences obtained from pronase-treated PHFs have shown that both three and four repeats contribute to the core of the PHF but that a length of protein containing only three repeats is protected (Jakes et al. 1991). This indicates that the amino-terminal half of tau is lost by proteolysis either in situ, because of endogenous proteases, or during tangle purification, because of added proteases. In the Alzheimer's disease brain, the amino-terminal half and part of the carboxy-terminus of tau are lost during the transition from intra- to extracellular tangles (Bondareff et al. 1990).

The above results were obtained using PHFs extracted from tangle fragments. Because of the insolubility of these filaments, it was not possible to conclude that tau is the only component of the PHF or to show whether there are other components. This situation has changed with the development of extraction techniques for dispersed PHFs (Greenberg and Davies 1990). A study involving purification of these filaments to apparent homogeneity has provided strong evidence that tau forms the PHF (Lee et al. 1991).

In addition, it has been shown that paired helical-like filaments can be assembled in vitro from bacterially expressed non-phosphorylated three-repeat or four-repeat fragments of tau (Fig. 10; Wille et al. 1992; Crowther et al. 1992). Filaments similar to PHFs were produced by dialysis or by hanging drop equilibration of the fragments against high concentrations of Tris at acidic pH. A series of filaments made from three repeats of tau is shown in Fig. 10b–e, compared with a pronase-treated PHF from an Alzheimer tangle preparation (Fig. 10a). The dimensions of the artificial filaments are very much like those of Alzheimer PHFs (Fig. 10b–e). A characteristic pattern of four longitudinal white, stain-excluding lines can also be seen. The production of artificial paired

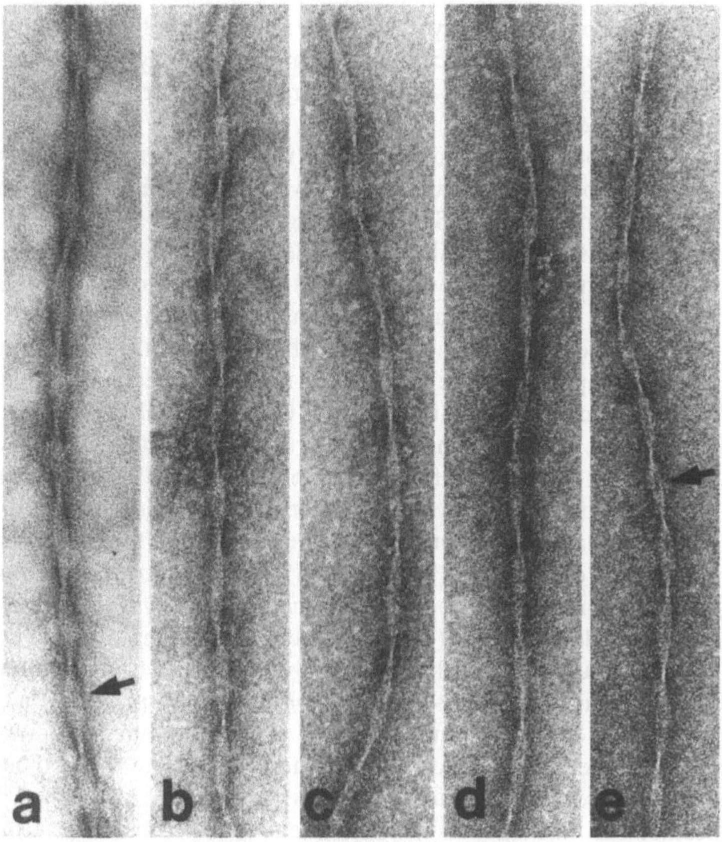

Fig. 10a–e. Paired helical filament (PHF) from the brain of an Alzheimer's disease patient and filaments assembled in vitro from a non-phosphorylated 99 amino acid fragment of tau encompassing three tandem repeats **a** Pronase-treated Alzheimer's disease PHF. **b–e** Filaments resembling PHFs assembled from expressed tau protein. The *arrows* indicate regions of filament where the characteristic pattern of four white stain-excluding lines parallel to the filament axis can be seen

helical-like filaments lends strong support to the view that tau is the only component necessary to form the PHF.

Unlike tangle fragment PHFs, a large proportion of dispersed PHFs contains the whole of tau (Lee et al. 1991; Ksiezak-Reding and Yen 1991; Brion et al. 1991; Goedert et al. 1992a). This indicates that dispersed filaments constitute an earlier stage of PHFs than the bulk of filaments isolated from tangle fragments. A further difference between tangle fragment and dispersed PHFs is that the majority of tangle fragment PHFs are ubiquitinated (Mori et al. 1987; Perry et al. 1987). Several lysine residues in the repeat region of tau that ubiquitin becomes attached to have been identified, indicating that these residues are located on the surface of the PHF (Morishima-Kawashima et al.

1993). The natural history of the PHF after assembly thus leads from an initial guanidine-soluble intracellular form to a very insoluble, proteolysed, ubiquitinated and cross-linked extracellular form, with a multitude of intermediates.

Tau protein extracted from PHFs runs as three major bands of 60-, 64- and 68-kDa apparent molecular mass, with a variable amount of background smear (Flament et al. 1989; Greenberg and Davies 1990; Lee et al. 1991). The latter results from partially proteolysed and cross-linked PHFs. These PHF-tau bands run more slowly on gels than normal or recombinant tau. They contain the whole of tau, as they stain with antibodies directed against the amino- and carboxy-termini of tau. After treatment with alkaline phosphatase at high temperature, the three PHF-tau bands become six bands that align with the recombinant tau isoforms (Goedert et al. 1992a; Greenberg et al. 1992; Liu et al. 1993). The relative proportions of the six bands are the same as those of soluble tau from adult brain. Thus, PHFs contain all six adult brain tau isoforms in a hyperphosphorylated state. As expected, brain regions with abundant neurofibrillary lesions contain large amounts of PHF-tau (Bramblett et al. 1992; Holzer et al. 1994). Moreover, the phosphorylation-dependent anti-tau antibodies stain neurofibrillary lesions in a phosphorylation-dependent manner (Grundke-lqbal et al. 1986; Mercken et al. 1992).

PHF-tau has a greatly reduced ability to bind to microtubules; this functional impairment results entirely from hyperphosphorylation, since dephosphorylated PHF-tau binds as well to microtubules as does normal tau (Bramblett et al. 1993; Yoshida and Ihara 1993). The reduced binding of PHF-tau to microtubules, coupled with reduced levels of normal tau, probably destabilises microtubules in Alzheimer's disease, resulting in the impairment of vital cellular processes, such as rapid axonal transport, and leading to the degeneration of affected nerve cells.

PHF-tau is both abnormally phosphorylated and hyperphosphorylated (Fig. 9). Abnormal phosphorylation means that more residues are phosphorylated in PHF-tau than in normal brain tau; hyperphosphorylation means that, for a given site, a higher proportion of tau molecules is phosphorylated in PHF-tau than in normal brain tau. To date, a large number of these sites have been identified by mass spectrometry (Hasegawa et al. 1992; Morishima-Kawashima et al. 1995) and through the use of phosphorylation-dependent anti-tau antibodies (Lee et al. 1991; Otvos et al. 1994; Goedert et al. 1994b, 1995; Seubert et al. 1995). They include serines 198, 199, 202, 208, 210, 214, 231, 235, 262, 396, 400, 404, 409, 412, 413 and 422 and threonines 181, 205, 212, 217 and 403 (Fig. 9). This shows that, besides the residues that are phosphorylated in tau from normal adult brain, a number of additional sites are phosphorylated in PHF-tau. Some of the latter are also phosphorylated in tau from newborn brain. About half of the known phosphorylation sites are serine and threonine residues followed by a proline. With the exception of Ser262, all known sites are located outside the microtubule-binding repeat region.

These findings imply the deregulation of phosphorylation-dephosphorylation mechanisms in Alzheimer's disease. MAP kinase, GSK3 and cdk5 all phosphorylate some of the hyperphosphorylated Ser/Thr-Pro sites of tau in vitro (Drewes et al. 1992; Ledesma et al. 1992; Goedert et al. 1992c, 1994b; Hanger et al. 1992; Mandelkow et al. 1992; Ishiguro et al. 1993; Paudel et al. 1993; Kobayashi et al. 1993; Baumann et al. 1993). Moreover, a cdc2-like enzyme has been found to be intimately associated with PHFs from Alzheimer's disease brain (Liu et al. 1995). Cyclic AMP-dependent protein kinase phosphorylates three of the non-Ser/Thr-Pro sites (S214, S262 and S409), while it also phosphorylates another three sites (S324, S356 and S416) that have not been reported to be phosphorylated in PHF-tau (Scott et al. 1993; J.M. Litersky et al., submitted for publication). Calcium/calmodulin-dependent protein kinase II phosphorylates one non-Ser/Thr-Pro site (S262) that is phosphorylated in PHF-tau and one site (S416) that has not been reported to be phosphorylated (Steiner et al. 1990; J.M. Litersky et al., submitted). However, it is at present unclear whether the underlying defect results from an increased protein kinase activity, a decreased protein phosphatase activity or a combination of both.

Whatever the mechanisms involved, it appears unlikely that tau is the only protein that is hyperphosphorylated. In affected nerve cells, PHFs probably constitute the most visible manifestation of the deregulation of phosphorylation/dephosphorylation mechanisms at Ser/Thr-Pro sites and non-Ser/Thr-Pro sites. Normal brain function might be disrupted by the hyperphosphorylation of proteins that are not normally phosphorylated to such an extent in vivo.

The degree of phosphorylation is the only known difference between PHF-tau and normal adult tau, with almost all the hyperphosphorylated sites located outside the tandem repeat region. As a consequence, tau is prevented from binding to microtubules and is believed to self-assemble through the tandem repeat region into the PHF.

Outlook

The molecular nature of the PHF has only recently been understood, with some hints as to what may lead to its formation. However, at present, we only have a very partial understanding of the mechanisms that lead to the phosphorylation of tau in normal brain and to its hyperphosphorylation in Alzheimer's disease. The majority of phosphorylated sites in normal brain tau and PHF-tau have been identified by mass spectrometry and the use of phosphorylation-dependent anti-tau antibodies. This has provided the necessary baseline for identifying protein kinases and protein phosphatases that regulate the phosphorylation state of tau. Thus, several candidate enzymes have been described that phosphorylate tau at some of the relevant sites in vitro. Similarly, apolipoprotein E has been shown to bind to recombinant tau, but not to

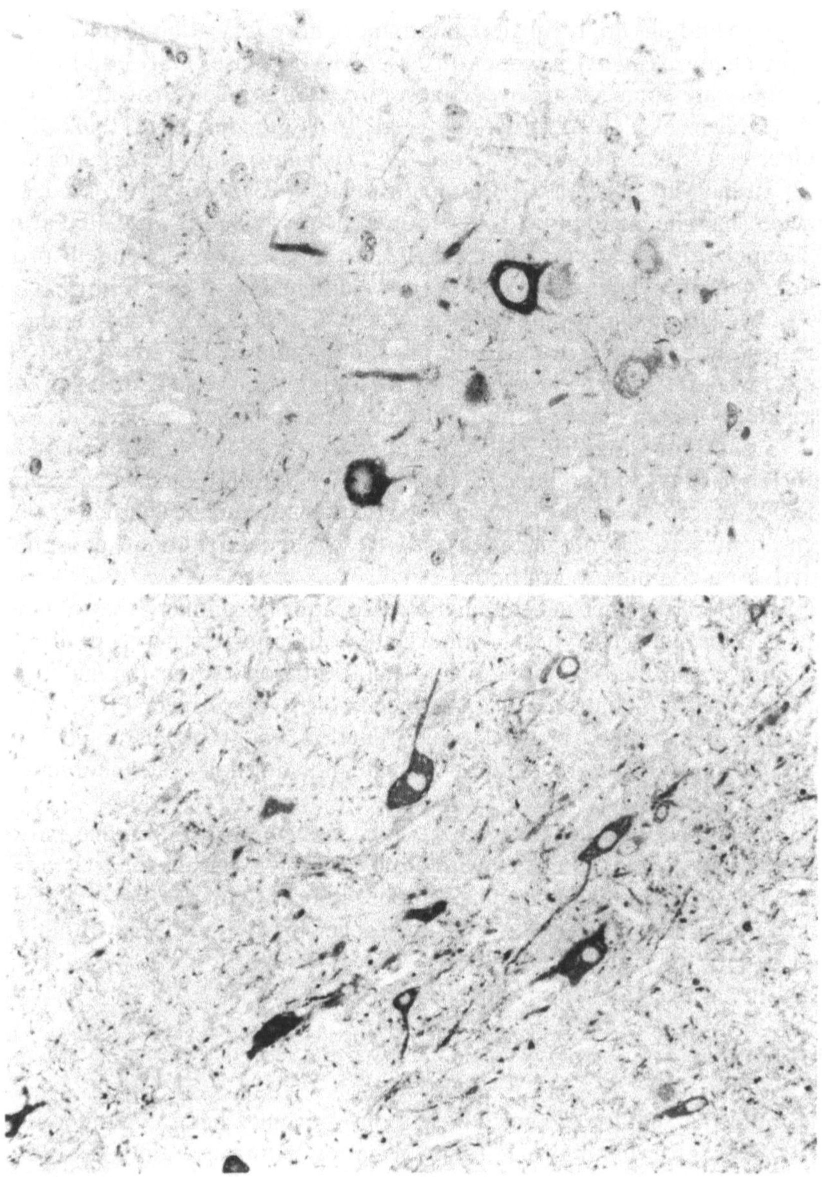

Fig. 11. Immunohistochemical staining of human tau in the brain of a mouse transgenic for the longest human brain tau isoform The anti-tau serum 133 was used *Top*, section through the cerebral cortex *Bottom*, section through the brain stem Note the strong somatodendritic staining of some nerve cells, it resembles the pre-tangle lesions of Alzheimer's disease

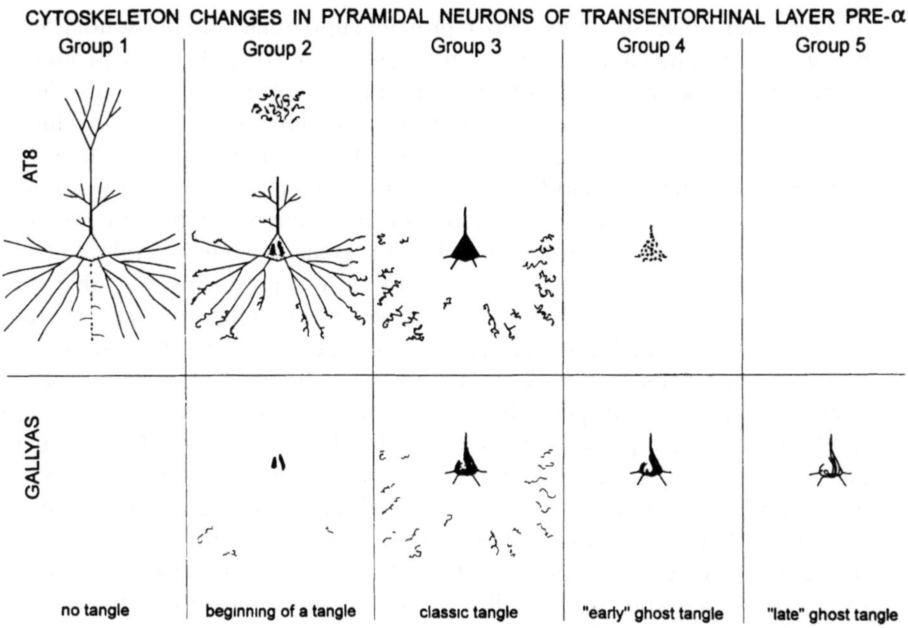

Fig. 12. Diagram comparing immunostaining with the phosphorylation-dependent anti-tau antibody AT8 and the Gallyas silver staining during development of neurofibrillary lesions. The progression of pathological alterations of the neuronal cytoskeleton is shown from the group 1 neuron to the group 5 structure. *Fine dots* indicate granular AT8 immunostaining, while *large dots* represent degenerating terminals that are attached to the disintegrated cell body

phosphorylated tau in vitro, suggesting that it may modulate the rate at which tau becomes phosphorylated in Alzheimer's disease brain (Strittmatter et al. 1994b). However, all these studies suffer from the inherent limitation that in vitro studies are not necessarily representative of the in vivo situation.

The next step must therefore be to relate these in vitro findings to the in vivo situation. At first sight one way to achieve this would be to compare protein kinase and protein phosphatase activities in Alzheimer's disease brain with those in control brain. Unfortunately, such an undertaking is hampered by the limited postmortem stability of protein kinase and protein phosphatase activities. An additional drawback is that, in a given piece of brain, the percentage of cells affected by the neurofibrillary pathology is only a fraction of those expressing the various enzymes, thereby rendering doubtful the interpretation of tissue homogenate measurements of enzyme activities. One way forward may be to use the information gained from in vitro studies to induce hyperphosphorylation of tau in intact cells and in animal brain. Studies on intact cells have already shown that over-expression of GSK3 results in hyperphosphorylation of tau (Lovestone et al. 1994; Sperber et al. 1995).

Transgenic animal technology and gene targeting are likely to play an increasingly important role in the investigation into the neurofibrillary lesions of Alzheimer's disease. One could in principle upregulate tau phosphorylation by over-expressing candidate protein kinases in nerve cells or by producing animals that lack protein phosphatases that are involved in the dephosphorylation of tau. We have taken a first step in that direction by producing transgenic mice that express the longest human brain tau isoform under control of the human Thy1 promoter (Götz et al. 1995). These animals show prominent axonal and somatodendritic staining of human tau in a subset of nerve cells in a number of brain regions, including cerebral cortex and hippocampal formation (Fig. 10), thereby reproducing the abnormal cellular localisation of tau that is observed in Alzheimer's disease brain. Moreover, transgenic human tau is phosphorylated at some of the same sites that are hyperphosphorylated in PHF-tau. Although PHFs have not yet been detected, the transgenic human tau protein shows pre-tangle changes similar to those that precede the full neurofibrillary pathology of Alzheimer's disease. The earliest known change in tau in the events leading to the neurofibrillary lesions of Alzheimer's disease is the appearance of hyperphosphorylated tau protein within individual nerve cells in the transentorhinal cortex (Figs. 11 and 12; Braak et al. 1994). The hope is that these and similar approaches will result in the development of a true animal model for the neurofibrillary lesions, which is an essential requirement for the testing of compounds aimed at halting or preventing the intracellular pathology of Alzheimer's disease.

Acknowledgments. I am grateful to Dr. D. Dickson for Fig. 4 and Dr. H. Braak for Fig. 11.

References

Arriagada PV, Growdon JH, Hedley-White ET, Hyman BT (1992) Neurofibrillary tangles but not senile plaques parallel duration and severity of Alzheimer's disease Neurology 42 631–639

Baumann K, Mandelkow EM, Biernat J, Piwnica-Worms H, Mandelkow E (1993) Abnormal Alzheimer's-like phosphorylation of tau protein by cyclin-dependent kinases cdk2 and cdk5 FEBS Lett 336 417–424

Binder LI, Frankfurter A, Rebhun LI (1985) The distribution of tau in the mammalian central nervous system J Cell Biol 101 1371–1378

Bondareff W, Wischik CM, Novak M, Amos WB, Klug A, Roth M (1990) Molecular analysis of neurofibrillary degeneration in Alzheimer's disease an immunohistochemical study Am J Pathol 37 711–723

Braak H, Braak E (1991) Neuropathological stageing of Alzheimer-related changes Acta Neuropathol 82 239–259

Braak E, Braak H, Mandelkow EM (1994) A sequence of cytoskeleton changes related to the formation of neurofibrillary tangles and neuropil threads Acta Neuropathol 87 554–567

Bramblett GT, Trojanowski JQ, Lee VM-Y (1992) Regions with abundant neurofibrillary pathology in human brain exhibit a selective reduction in levels of binding-competent tau and accumulation of abnormal tau isoforms (A68 proteins) Lab Invest 66· 212–222

Bramblett GT, Goedert M, Jakes R, Merrick SE, Trojanowski JQ, Lee VM-Y (1993) Abnormal tau phosphorylation at Ser^{396} in Alzheimer's disease recapitulates development and contributes to reduced microtubule binding Neuron 10 1089–1099

Brion JP, Passareiro H, Nunez J, Flament-Durand J (1985) Mise en évidence immunologique de la protéine tau au niveau des lésions de dégénérescence neurofibrillaire de la maladie d'Alzheimer. Arch Biol (Bruxelles) 95 229–235

Brion JP, Hanger DP, Bruce MT, Couck AM, Flament-Durand J, Anderton BH (1991) Tau in Alzheimer neurofibrillary tangles: N-and C-terminal regions are differentially associated with paired helical filaments and the location of a putative abnormal phosphorylation site. Biochem J 273: 127–133

Brion JP, Smith C, Couck AM, Gallo JM, Anderton BH (1993) Developmental changes in tau phosphorylation. fetal tau is transiently phosphorylated in a manner similar to paired helical filament tau characteristic of Alzheimer's disease J Neurochem 61. 2071–2080

Butler M, Shelanski ML (1986) Microheterogeneity of microtubule-associated tau protein is due to differences in phosphorylation J Neurochem 47: 1517–1522

Butner KA, Kirschner MW (1991) Tau protein binds to microtubules through a flexible array of distributed weak sites J Cell Biol 115 717–730

Caceres A, Kosik KS (1990) Inhibition of neurite polarity by tau antisense oligonucleotides in primary cerebellar neurons Nature 343 461–463

Couchie D, Mavlia C, Georgieff IS, Liem RKH, Shelanski ML, Nunez J (1992) Primary structure of high molecular weight tau present in the peripheral nervous system Proc Natl Acad Sci USA 89 4378–4381

Crowther RA (1991) Straight and paired helical filaments in Alzheimer disease have a common structural unit Proc Natl Acad Sci USA 88· 2288–2292

Crowther RA, Wischik CM (1985) Image reconstruction of the Alzheimer paired helical filament EMBO J 4 3661–3665

Crowther RA, Olesen OF, Jakes R, Goedert M (1992) The microtubule binding repeats of tau protein assemble into filaments like those found in Alzheimer's disease FEBS Lett 309 199–202

Dickson DW, Crystal HA, Mattiace LA, Masur DM, Blau AD, Davies P, Yen S-H, Aronson MK (1991) Identification of normal and pathological aging in prospectively studied non-demented elderly humans Neurobiol Aging 13 179–189

Drechsel DN, Hyman AA, Cobb MH, Kirschner MW (1992) Modulation of the dynamic instability of tubulin assembly by the microtubule-associated protein tau. Mol Cell Biol 3· 1141–1154

Drewes G, Lichtenberg-Kraag B, Doring F, Mandelkow EM, Biernat J, Dorée M, Mandelkow E (1992) Mitogen-activated protein (MAP) kinase transforms tau protein into an Alzheimer-like state EMBO J 11 2131–2138

Drewes G, Trinczek B, Illenberger S, Biernat J, Schmitt-Ulms G, Meyer HE, Mandelkow EM, Mandelkow E (1995) Microtubule-associated protein/microtubule affinity-regulating kinase (p110 mark) J Biol Chem 270 7679–7688

Drubin DG, Kirschner MW (1986) Tau protein function in living cells J Cell Biol 103 2739–2746

Edson K, Weisshaar B, Matus A (1993) Actin depolymerisation induces process formation in MAP2-transfected neuronal cells Development 117 689–700

Flament S, Delacourte A, Hémon B, Défossez A (1989) Characterization of two pathological tau protein variants in Alzheimer's disease J Neurol Sci 92· 133–141

Gallo JM, Hanger DP, Twist EC, Kosik KS, Anderton BH (1992) Expression and phosphorylation of a three-repeat isoform of tau in transfected non-neuronal cells Biochem J 286 399–404

Glenner G, Wong CW (1984) Alzheimer's disease initial report of the purification and characterization of a novel cerebrovascular amyloid protein Biochem Biophys Res Commun 120: 885–890

Goedert M (1993) Tau protein and the neurofibrillary pathology of Alzheimer's disease Trends Neurosci 16: 460–465

Goedert M, Jakes R (1990) Expression of separate isoforms of human tau protein· correlation with the tau pattern in brain and effects on tubulin polymerization. EMBO J 9 4225–4230

Goedert M, Wischik CM, Crowther RA, Walker JE, Klug A (1988) Cloning and sequencing of the cDNA encoding a core protein of the paired helical filament of Alzheimer disease Proc Natl Acad Sci USA 85 4051–4055

Goedert M, Spillantini MG, Jakes R, Rutherford D, Crowther RA (1989a) Multiple isoforms of human microtubule-associated protein tau sequences and localization in neurofibrillary tangles of Alzheimer's disease Neuron 3. 519–526

Goedert M, Spillantini MG, Potier MC, Ulrich J, Crowther RA (1989b) Cloning and sequencing of the cDNA encoding an isoform of microtubule-associated protein tau containing four tandem repeats differential expression of tau protein mRNAs in human brain EMBO J 8 393–399

Goedert M, Spillantini MG, Cairns NJ, Crowther RA (1992a) Tau proteins of Alzheimer paired helical filaments abnormal phosphorylation of all six brain isoforms Neuron 8 159–168

Goedert M, Spillantini MG, Crowther RA (1992b) Cloning of a big tau microtubule-associated protein characteristic of the peripheral nervous system. Proc Natl Acad Sci USA 89 1983–1987

Goedert M, Cohen ES, Jakes R, Cohen P (1992c) p42 MAP kinase phosphorylation sites in microtubule-associated protein tau are dephosphorylated by protein phosphatase $2A_1$ Implications for Alzheimer's disease FEBS Lett 312 95–99

Goedert M, Jakes R, Crowther RA, Six J, Lubke U, Vandermeeren M, Cras P, Trojanowski JQ, Lee VM-Y (1993) The abnormal phosphorylation of tau protein at serine 202 in Alzheimer disease recapitulates phosphorylation during development Proc Natl Acad Sci USA 90 5066–5070

Goedert M, Jakes R, Spillantini MG, Crowther RA (1994a) Tau protein and Alzheimer's disease In. Hyams JS, Lloyd CW (eds) Microtubules, Wiley-Liss Inc, pp 183–200

Goedert M, Jakes R, Crowther RA, Cohen P, Vanmechelen E, Vandermeeren M, Cras P (1994b) Epitope mapping of monoclonal antibodies to the paired helical filaments of Alzheimer's disease identification of phosphorylation sites in tau protein Biochem J 301. 871–877

Goedert M, Jakes R, Vanmechelen E (1995) Monoclonal antibody AT8 recognises tau protein phosphorylated at both serine 202 and threonine 205 Neurosci Lett 189 167–170

Gotz J, Probst A, Spillantini MG, Schafer T, Jakes R, Burki K, Goedert M (1995) Somatodendritic localisation and hyperphosphorylation of tau protein in transgenic mice expressing the longest human brain tau isoform EMBO J 4 1304–1313

Goode BL, Feinstein SC (1994) Identification of a novel microtubule binding and assembly domain in the developmentally regulated inter-repeat region of tau J Cell Biol 124 769–782

Greenberg SG, Davies P (1990) A preparation of Alzheimer paired helical filaments that displays distinct tau proteins by polyacrylamide gel electrophoresis Proc Natl Acad Sci USA 87 5827–5831

Greenberg SG, Davies P, Schein JD, Binder LI (1992) Hydrofluoric acid-treated tau PHF proteins display the same biochemical properties as normal tau. J Biol Chem 267 564–569

Grundke-Iqbal I, Iqbal K, Tung Y-C, Zaidi MS, Wisniewski HM, Binder LI (1986) Abnormal phosphorylation of the microtubule-associated protein tau in Alzheimer cytoskeletal pathology. Proc Natl Acad Sci USA 83: 4913–4917

Hanemaaijer R, Ginzburg I (1991) Involvement of mature tau isoforms in the stabilization of neurites in PC12 cells J Neurosci Res 30 163–171

Hanger DP, Hughes K, Woodgett JR, Brion JP, Anderton BH (1992) Glycogen synthase kinase-3 induces Alzheimer's disease-like phosphorylation of tau generation of paired helical filament epitopes and neuronal localization of the kinase. Neurosci Lett 147: 58–62

Harada A, Oguchi K, Okabe S, Kuno J, Terada S, Ohshima T, Sato-Yoshitake R, Takei Y, Noda T, Hirokawa N (1994) Altered microtubule organization in small-calibre axons of mice lacking tau protein Nature 369 488–491

Hasegawa M, Morishima-Kawashima M, Takio K, Suzuki M, Titani K, Ihara Y (1992) Protein sequence and mass spectrometric analyses of tau in the Alzheimer's disease brain J Biol Chem 267 17047–17054

Hasegawa M, Watanabe A, Takio K, Suzuki M, Arai T, Titani K, Ihara Y (1993) Characterization of two distinct monoclonal antibodies to paired helical filaments· further evidence for fetal-type phosphorylation of the tau in paired helical filaments J Neurochem 60 2068–2077

Hirokawa N, Shiomura Y, Ogabe S (1988) Tau proteins The molecular structure and mode of binding on microtubules J Cell Biol 107. 1449–1459

Holzer M, Holzapfel HP, Zedlick D, Bruckner MK, Arendt T (1994) Abnormally phosphorylated tau protein in Alzheimer's disease· heterogeneity of individual regional distribution and relationship to clinical severity Neuroscience 63 499–516

Ishiguro K, Shiratsuchi A, Sato S, Omori A, Arioka M, Kobayashi S, Uchida T, Imahori K (1993) Glycogen synthase kinase 3β is identical to tau protein kinase I generating several epitopes of paired helical filaments FEBS Lett 325 167–172

Jakes R, Novak M, Davison M, Wischik CM (1991) Identification of 3- and 4-repeat tau isoforms within the PHF in Alzheimer's disease EMBO J 10 2725–2729

Kanai Y, Takemura R, Oshima T, Mori H, Ihara Y, Yanagisawa M, Masaki T, Hirokawa N (1989) Expression of multiple tau isoforms and microtubule bundle formation in fibroblasts transfected with a single tau cDNA J Cell Biol 109 1173–1184

Kanai Y, Chen J, Hirokawa N (1992) Microtubule bundling by tau proteins in vivo Analysis of functional domains EMBO J 11 3953–3961

Kanemaru K, Takio K, Miura R, Titani K, Ihara Y (1992) Fetal-type phosphorylation of the tau in paired helical filaments J Neurochem 58 1667–1675

Kenessey A, Yen S-HC (1993) The extent of phosphorylation of fetal tau is comparable to that of PHF-tau from Alzheimer paired helical filaments Brain Res 629· 40–46

Kidd M (1963) Paired helical filaments in electron microscopy of Alzheimer's disease Nature 197 192–193

Knops J, Kosik KS, Lee G, Pardee JD, Cohen-Gould L, McColongue L (1991) Overexpression of tau in a nonneuronal cell induces long cellular processes J Cell Biol 114 725–733

Kobayashi S, Ishiguro K, Omori A, Takamatsu M, Arioka M, Imahori K, Uchida T (1993) A cdc2-related kinase PSSALRE/cdk5 is homologous with the 30 kDa subunit of tau protein kinase II, a proline-directed protein kinase associated with microtubules FEBS Lett 335. 171–175

Kondo J, Honda T, Mori H, Hamada Y, Miura R, Ogawara H, Ihara Y (1988) The carboxyl third of tau is tightly bound to paired helical filaments Neuron 1. 827–834

Kowall NW, Beal MF, Busciglio J, Duffy LK, Yankner BA (1991) An in vivo model for the neurodegenerative effects of β-amyloid and protection by substance P Proc Natl Acad Sci USA 88: 7247–7251

Ksiezak-Reding H, Yen S-H (1991) Structural stability of paired helical filaments requires microtubule-binding domains of tau a model for self-association Neuron 6: 717–728

Ledesma MD, Correas I, Avila J, Diaz-Nido J (1992) Implication of brain cdc2 and MAP2 kinases in the phosphorylation of tau in Alzheimer's disease. FEBS Lett 308· 218–224

Lee G, Rook SL (1992) Expression of tau protein in non-neuronal cells. Microtubule binding and stabilization J Cell Sci 102: 227–237

Lee G, Cowan N, Kirschner M (1988) The primary structure and heterogeneity of tau protein from mouse brain Science 239· 285–288

Lee VM-Y, Balin BJ, Otvos L, Trojanowski JQ (1991) A68 – a major subunit of paired helical filaments and derivatized forms of normal tau. Science 251 675–678

Liu W-K, Dickson DW, Yen S-H (1993) Heterogeneity of tau proteins in Alzheimer's disease Am J Pathol 142: 387–394

Liu W-K, Williams RT, Hall FL, Dickson DW, Yen S-H (1995) Detection of a cdc2-related kinase associated with Alzheimer paired helical filaments Am J Pathol 146 228–238

Lo MMS, Fieles AW, Norris TE, Dargis DG, Caputo CB, Scott CW, Lee VM-Y, Goedert M (1993) Human tau isoforms confer distinct morphological and functional properties to stably transfected fibroblasts Mol Brain Res 20: 209–220

Lovestone S, Reynolds CH, Latimer D, Davis DR, Anderton BH, Gallo JM, Hanger D, Mulot S, Marquardt B, Stabel S, Woodgett JR, Miller CCJ (1994) Alzheimer's disease-like phosphorylation of the microtubule-associated protein tau by glycogen synthase kinase-3 in transfected mammalian cells Curr Biol 4 1077–1086

Mandelkow EM, Drewes G, Biernat J, Gustke N, Van Lint J, Vandenheede JR, Mandelkow E (1992) Glycogen synthase kinase-3 and the Alzheimer-like state of microtubule-associated protein tau FEBS Lett 314 315–321

Masters CL, Simms G, Weinman NA, Multhaup G, McDonald BL, Beyreuther K (1985) Amyloid plaque core protein in Alzheimer's disease and Down's syndrome Proc Natl Acad Sci USA 82 4245–4249

Matsuo ES, Shin R-W, Bilingsley ML, Van de Voorde A, O'Connor M, Trojanowski JQ, Lee VM-Y (1994) Biopsy-derived adult human brain tau is phosphorylated at many of the same sites as Alzheimer's disease paired helical filament tau Neuron 13 989–1002

Mercken M, Vandermeeren M, Lubke U, Six J, Boons J, Van de Voorde A, Martin JJ, Gheuens J (1992) Monoclonal antibodies with selective specificity for Alzheimer tau are directed against phosphatase-sensitive epitopes Acta Neuropathol 84: 265–272

Mori H, Kondo J, Ihara Y (1987) Ubiquitin is a component of paired helical filaments in Alzheimer's disease Science 35 1641–1644

Morishima-Kawashima M, Hasegawa M, Takio K, Suzuki M, Titani K, Ihara Y (1993) Ubiquitin is conjugated with amino-terminally processed tau in paired helical filaments Neuron 10 1151–1160

Morishima-Kawashima M, Hasegawa M, Takio K, Suzuki M, Yoshida H, Titani K, Ihara Y (1995) Proline-directed and non-proline-directed phosphorylation of PHF-tau. J Biol Chem 270 823–829

Mullan M, Crawford F (1993) Genetic and molecular advances in Alzheimer's disease Trends Neurosci 16 398–403

Otvos L, Feiner L, Lang E, Szendrei GI, Goedert M, Lee VM-Y (1994) Monoclonal antibody PHF-1 recognizes tau protein phosphorylated at serine residues 396 and 404 J Neurosci Res 39 669–673

Paudel HK, Lew J, Zenobia A, Wang JH (1993) Brain proline-directed kinase phosphorylates tau on sites that are abnormally phosphorylated in tau associated with Alzheimer's paired helical filaments J Biol Chem 268 23512–23518

Peng I, Binder LI, Black MM (1986) Biochemical and immunological analyses of cytoskeletal domains of neurons J Cell Biol 102 252–262

Perry G, Friedman R, Shaw G, Chau V (1987) Ubiquitin is detected in neurofibrillary tangles and senile plaque neurites of Alzheimer disease brains Proc Natl Acad Sci USA 84 3033–3036

Poulter L, Barratt D, Scott CW, Caputo CB (1993) Locations and immunoreactivities of phosphorylation sites on bovine and porcine tau proteins and a PHF-tau fragment J Biol Chem 268 9636–9644

Scott CW, Spreen RC, Herman JL, Chow FP, Davison MD, Young J, Caputo CB (1993) Phosphorylation of recombinant tau by cAMP-dependent protein kinase J Biol Chem 268 1166–1173

Selkoe DJ (1994) Normal and abnormal biology of the β-amyloid precursor protein Annu Rev Neurosci 17 489–517

Seubert P, Mawal-Dewan M, Barbour R, Jakes R, Goedert M, Johnson GVW, Litersky JM, Schenk D, Lieberburg I, Trojanowski JQ, Lee VM-Y (1995) Detection of phosphorylated Ser^{262} in fetal tau, adult tau and paired helical filament tau J Biol Chem 270 18917–18922

Sontag E, Nunbhakdi-Craig V, Bloom GS, Mumby MC (1995) A novel pool of protein phosphatase 2A is associated with microtubules and is regulated during the cell cycle J Cell Biol 128 1131–1144

Sperber BR, Leight S, Goedert M, Lee VM-Y (1995) Glycogen synthase kinase-3β phosphorylates tau protein at multiple sites in intact cells Neurosci Lett 197 149–153

Steiner B, Mandelkow EM, Biernat J, Gustke N, Meyer HE, Schmidt B, Mieskes G, Soling HD, Drechsel D, Kirschner MW, Goedert M, Mandelkow E (1990) Phosphorylation of microtubule-associated protein tau: identification of the site for Ca^{2+}-calmodulin dependent kinase and relationship with tau phosphorylation in Alzheimer tangles EMBO J 9 3539–3544

Strittmatter WJ, Weisgraber KH, Goedert M, Saunders AM, Huang D, Corder EH, Dong L-M, Jakes R, Alberts MJ, Gilbert JR, Han S-H, Hulette C, Einstein G, Schmechel DE, Pericak-Vance MA, Roses AD (1994a) Hypothesis· microtubule instability and paired helical filament formation in the Alzheimer disease brain are related to apolipoprotein E genotype Exp Neurol 125 163–171

Strittmatter WJ, Saunders AM, Goedert M, Weisgraber KH, Dong L-M, Jakes R, Huang DY, Pericak-Vance M, Schmechel D, Roses AD (1994b) Isoform-specific interactions of apolipoprotein E with microtubule-associated protein tau implications for Alzheimer disease Proc Natl Acad Sci USA 91 11183–11186

Szucs K, Ledesma MD, Dombradi V, Gergely P, Avila J, Friedrich P (1994) Dephosphorylation of tau protein from Alzheimer's disease patients Neurosci Lett 165 175–178

Watanabe A, Hasegawa M, Suzuki M, Takio K, Morishima-Kawashima M, Titani K, Arai T, Kosik KS, Ihara Y (1993) In vivo phosphorylation sites in fetal and adult rat tau J Biol Chem 268 25712–25717

Weingarten MD, Lockwood AH, Hwo S-H, Kirschner MW (1975) A protein factor essential for microtubule assembly Proc Natl Acad Sci USA 72 1858–1862

Wille H, Drewes G, Biernat J, Mandelkow EM, Mandelkow E (1992) Alzheimer-like paired helical filaments and antiparallel dimers formed from microtubule-associated protein tau in vitro J Cell Biol 118 573–584

Wischik CM, Novak M, Thogersen HC, Edwards PC, Runswick MJ, Jakes R, Walker JE, Milstein C, Roth M, Klug A (1988) Isolation of a fragment of tau derived from the core of the paired helical filament of Alzheimer disease Proc Natl Acad Sci USA 85 4506–4510

Yoshida H, Ihara Y (1993) Tau in paired helical filament is functionally distinct from fetal tau assembly incompetence of paired helical filament tau J Neurochem 61 1183–1186

Tau Protein and Alzheimer Paired Helical Filament Assembly: Interacting Domains and Control Residues

E. Mandelkow, O. Schweers, J. Biernat, B. Trinczek, G. Drewes,* and *E.-M. Mandelkow*

Abstract

Tau protein can interact with tubulin, the subunits of microtubules, and it can interact with itself in the paired helical filaments of Alzheimer's disease. Both interactions are thought to be caused by post-translational modifications of the protein. We have analyzed the domain structure of tau, the effect of the domains on tau-tubulin or tau-tau interactions, and the residues that are critically involved. Two key residues appear to be particularly important in switching between different states, Ser-262 and Cys-322. Ser-262 can be phosphorylated by the kinase p110(MARK), which abolishes tau's binding to microtubules. Cys-322 can be oxidized to disulfide bridges, which leads to the formation of antiparallel tau dimers and paired helical filaments. The results argue that the tau-microtubule interactions and the tau-tau interactions leading to PHFs are controlled by different chemical modifications.

Introduction

Tau protein is a neuronal microtubule-associated protein. Its functions include the stabilization of microtubules, the outgrowth of neurites, and the spacing between microtubules and their neighbors (for a review, see Kosik and McConlogue 1994). Tau is also the main component of Alzheimer paired helical filaments (PHFs). Tau is a phosphoprotein, both in its physiological state (Butler and Shelanski 1986) and especially in PHFs (Grundke-Iqbal et al. 1986). One current hypothesis holds that tau's association with microtubules is regulated by (physiological) phosphorylation, whereas its detachment from microtubules and aggregation into PHFs is caused by (pathological) hyperphosphorylation. In a report to a previous IPSEN Foundation conference, we summarized the phosphorylation sites of tau, kinases acting on tau, and methods to analyze the "pathological" phosphorylation by means of diagnostic antibodies (Mandelkow et al. 1995a). Here we concentrate on the domains of tau that have to interact, either to stabilize microtubules or to

*Max-Planck-Unit for Structural Molecular Biology, c/o DESY, Notkestrasse 85, 22603 Hamburg, Germany

A D Roses et al (Eds)
Apolipoprotein E and Alzheimer's Disease

aggregate into PHFs, and the protein modifications that are involved (phosphorylation and oxidation).

Results and Discussion

Tau–Microtubule Interactions: The "Jaws" Model

Tau protein can be subdivided into several domains, based on the character of the amino acid sequence (Fig. 1). The N-terminal part is not involved in microtubule interactions; it probably projects away from the microtubule wall and need not concern us in the present context. Likewise, the C-terminal tail is not involved in microtubule binding. This leaves the domain of the repeats (gray shading in Fig. 1) and the darkly shaded flanking regions P and R′ as the primary microtubule interacting regions (Gustke et al. 1994).

The repeat domain consists of three or four homologous 31-amino acid stretches, depending on tau isoform (the second repeat may be absent, especially in fetal tau; see Goedert et al. 1989). The repeats are sometimes referred to as the "microtubule-binding" domain. This is based on the observation that peptides or tau constructs comprising individual repeats or the repeat domain bind weakly to microtubules (with dissociation constants in the 0.2–1 m*M* range; Aizawa et al. 1988; Ennulat et al. 1989), and guided by the notion that a repetitive binding motif in tau would fit well with the repetitive nature of tubulin subunits in the microtubule wall (for a review of the lattice, see Mandelkow et al. 1995). However, this view is incomplete since the repeats alone do not explain the tight binding of tau to microtubules (K_d in the micromolar range for whole tau). The high affinity of tau-microtubule binding is obtained only when the flanking domains are present. In fact, it can be shown that repeat-less tau, which contains the flanking regions P and P′ but not

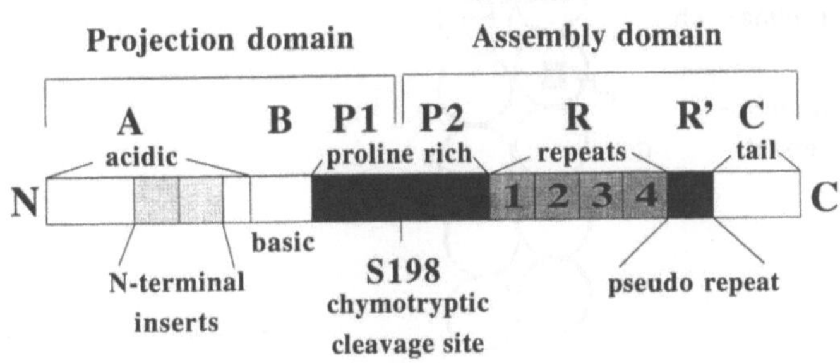

Fig. 1. Tau isoform htau40 (441 residues, Goedert et al 1989) The two 29-residue inserts near the N terminus are *lightly shaded*, and the four repeats in the C-terminal half are numbered 1–4 (*medium shade*) *Projection domain*, M1-Y197, does not bind to microtubules by itself, *assembly domain*, S198-L441, binds to microtubules, separable by chymotryptic cleavage (Gustke et al. 1994). N-terminal domains *A* (acidic, M1-G120), *B* (basic, G120-I151), *P* (proline-rich and basic I151-Q244), separated into *P1* and *P2* at Y197 Repeats R1-R4, Q244-N368, "fifth" repeat R′ K369-S400 *C*, C-terminal tail G401-L441

the repeats themselves, binds almost as tightly as whole tau (Gustke et al. 1994). This view is compatible with several studies demonstrating the strong binding of tau as long as the P-domain or parts of it are present (e.g., Butner and Kirschner 1991; Brandt and Lee 1993).

Strong binding of tau to microtubules may be necessary for regulating microtubule stability, but it need not be sufficient. We therefore asked how the binding of various tau constructs compared with their capacity to stabilize microtubules, nucleate their assembly, or support their elongation. The results can be described by distinguishing "productive" and "unproductive" binding, or by distinguishing "targetting" and "catalytic" domains (Fig. 2). Whole tau binds tightly, and it stabilizes microtubules efficiently. Repeat-less tau still binds strongly, but it is no longer productive because it does not promote microtubule stability (black parts of tau in Fig. 2). The repeats (white boxes) bind weakly by themselves, and they are unproductive as well because they have only a marginal influence on microtubule stability. These observations can be described by a "jaws" model. The flanking regions are viewed as targetting domains which fix tau on the surface of microtubules; they will do so even in the absence of the repeat domain (hence the tight binding). However, tau catalyzes microtubule assembly only when the repeats are also present.

Tau–Microtubule Interactions: The Role of Phosphorylation

If different domains of tau have different functions in the interaction with microtubules, then one would expect that regulation by phosphorylation might occur on different levels as well. This is indeed the case. The flanking

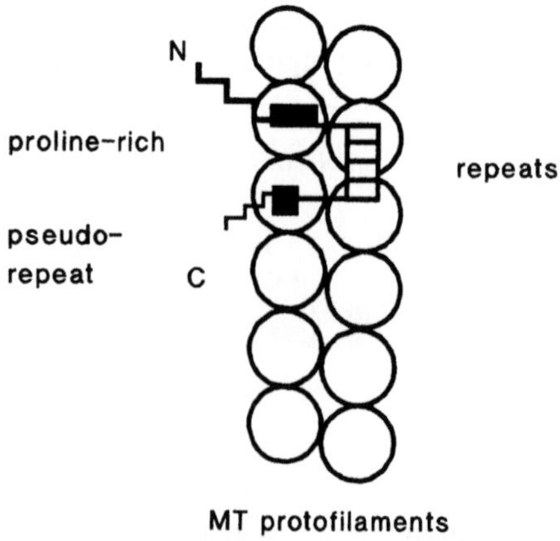

Fig. 2. "Jaws" model of tau–microtubule interactions The rows of *spheres* represent tubulin subunits in two protofilaments The targeting domains P and R′ (*black*) can bind tightly to one protofilament, but they do not stabilize interactions between protofilaments The repeat domain R also binds to a protofilament, albeit weakly, and its stabilization of bonds between protofilaments is equally weak. However, when targeting and catalytic domains are combined, the binding is strong and microtubules (*MT*) are stabilized

regions contain several Ser-Pro or Thr-Pro motifs that can be phosphorylated by several proline-directed kinases, such as MAP kinase, the neuronal cdc2-like kinase cdk5, or glycogen-synthase-kinase-3 (GSK3; for review, see Mandelkow et al. 1995). Phosphorylation of these motifs modulates the tau-microtubule interaction but does not abrogate it. By contrast, the repeats do not contain Ser/Thr-Pro motifs, which argues against a role of proline-directed kinases. However, there is a special residue, Ser-262, that can be phosphorylated by a novel kinase termed p110 (MARK), for "microtubule-affinity regulating kinase" (Drewes et al. 1995). Phosphorylation at this site causes a drastic reduction in the affinity of tau for microtubules and makes them dynamically unstable (Biernat et al. 1993).

Why are these sites important? It has long been noted that tau from Alzheimer PHFs is "hyperphosphorylated" and reacts with antibodies raised specifically against PHFs. A number of these "diagnostic" epitopes turned out to be phosphorylated Ser/Thr-Pro motifs in the flanking regions, suggesting that proline-directed kinases play a role in transforming tau protein into a "pathological state." Some of these antibodies allow one to map the progression of Alzheimer's disease at earlier stages than conventional silver staining methods (e.g., Braak et al. 1994). On the other hand, the importance of phosphorylated Ser-262 was highlighted by the finding that this residue was specifically phosphorylated in Alzheimer tau (Hasegawa et al. 1992). One interpretation of these results is that the phosphorylation of Ser-262 (e.g., by MARK) causes the detachment of tau from microtubules, leading to their decay and subsequent interruption of microtubule-based axonal transport pathways. Subsequently, tau becomes "hyperphosphorylated" and assembles into PHFs.

Assembly of Paired Helical Filaments: Role of Oxidation

Since tau in PHFs is both pathologically aggregated and hyperphosphorylated, it is often assumed that there is a link between these two features. However, it has been difficult to obtain direct evidence for this connection. PHF tau is modified in other ways as well, for example by ubiquitination (Morishima et al. 1993) and glycation (Yan et al. 1994), and oxidative damage is a common feature of aging brain tissue (Smith et al. 1991); all of these could potentially play a role in PHF aggregation. Our earlier attempts to reproduce this aggregation in vitro led to the following results. Whole tau, prepared either as recombinant protein or from brain tissue, shows a tendency to aggregate into filaments, but these do not display the characteristics expected of PHFs (Lichtenberg-Kraag and Mandelkow 1990; note that PHFs normally have widths between 10 and 20 nm and a cross-over repeat around 80 nm; Wischik et al. 1985). We found that constructs of tau containing essentially the repeat domain could be induced to form fibers with the morphological hallmarks of PHFs; this required that the constructs were chemically crosslinked into dimers so that the dimers appeared to be building blocks of PHFs in vitro

(Wille et al. 1992). The fact that the repeat domain is capable of assembling into PHFs in vitro agrees well with the observation that this same domain is also at the core of Alzheimer PHFs (Kondo et al. 1988; Wischik et al. 1988).

The polymerization of "amyloid" fibers is thought to be based on the interaction of protein domains with a beta-strand structure (Glenner et al. 1972) and, in the case of amyloid plaque fibers made up of the peptide A-beta, a cross-beta structure has indeed been demonstrated (Inouye et al. 1993). We sought to identify similar elements of secondary structure in tau; however, none of the spectroscopic and scattering methods employed (circular dichroism, Fourier-transform infrared spectroscopy, X-ray solution scattering and fiber diffraction) yielded signs of alpha-helix or beta-sheet conformations. In fact tau protein is best described as a random (Gaussian) polymer lacking a well-defined shape (Schweers et al. 1994). Thus, for the moment the question of how this highly hydrophilic and soluble protein precipitates into insoluble aggregates remains a mystery.

Next we attempted to demonstrate an influence of phosphorylation on PHF aggregation in vitro. These studies are still in progress and will be described elsewhere. However, another principle emerged from these studies: PHFs form most efficiently when the key residue Cys-322 is oxidized to form inter-molecular disulfide bridges (Schweers et al. 1995). The principle is illustrated in Fig. 3, in which we compare two constructs that have either three

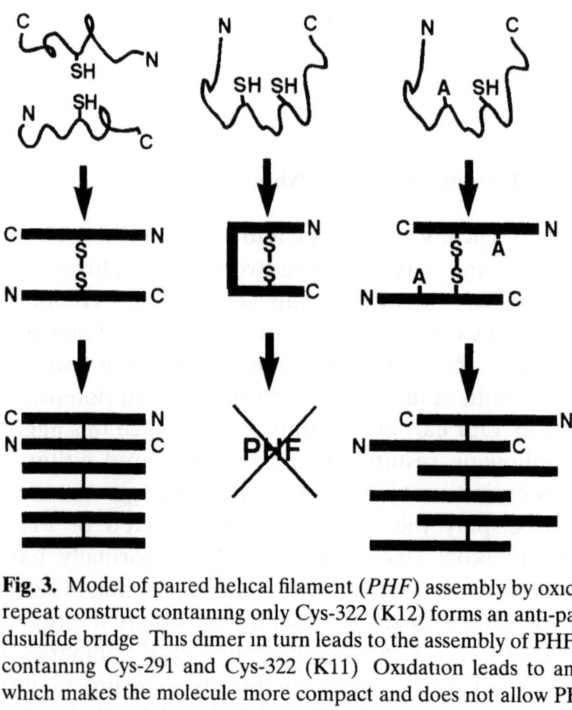

Fig. 3. Model of paired helical filament (*PHF*) assembly by oxidation of SH groups *Left*, a three-repeat construct containing only Cys-322 (K12) forms an anti-parallel dimer by an intermolecular disulfide bridge This dimer in turn leads to the assembly of PHFs *Middle*, a four-repeat construct containing Cys-291 and Cys-322 (K11) Oxidation leads to an intramolecular disulfide bridge, which makes the molecule more compact and does not allow PHF assembly *Right*, a four-repeat construct as in the middle row, but with Cys-291 mutated into Ala The single remaining Cys-322 is again available for intermolecular disulfide crossbridges and PHF formation

(K12, left row) or four repeats (K11, middle and right row). In the case of the three-repeat construct K12, oxidation of SH groups can only lead to intermolecular crossbridges because only the single Cys-322 in the third repeat is present (Fig. 3, left). The dimers can be visualized by native gel electrophoresis (not shown). Further aggregation of the dimers leads to PHF-like fibers similar to those observed previously with chemically cross-linked tau constructs (Wille et al. 1992). By contrast, the four-repeat construct K11 contains the two Cys residues Cys-291 and Cys-322 in the second and third repeats (Fig. 3, middle). Their oxidation could lead either to intra-molecular or inter-molecular crossbridges. In practice the intra-molecular disulfide bridge predominates; this leads to an internal stabilization and to a more compact form of the molecule. This can be demonstrated by native gel electrophoresis; the probable explanation is that the two Cys are in close proximity, even in the reduced state, so that pyrene residues attached to them show excimer fluorescence (data not shown). The important point is that in the internally oxidized four-repeat construct there are no more Cys available for the crosslinking of tau dimers, and consequently PHFs are no longer formed. This point can be reinforced by a control experiment where Cys-291 in the second repeat is mutated into Ala (Fig. 3, right). The mutant construct K11 (Cys-291-Ala) has four repeats but only the one Cys residue, Cys-322. This is now again available for inter-molecular disulfide crossbridges, and consequently the mutant construct can assemble into PHFs, similar to the three-repeat construct K12.

Summary

The observations described here indicate that phosphorylation is likely to play a role in regulating tau-microtubule interactions. However, a definitive role for phosphorylation in the tau-tau interaction leading to PHFs has not been demonstrated so far. By contrast, the initial states of PHF formation can be linked to the oxidation of SH groups; this applies in particular to the three-repeat form of tau. The data would be compatible with the following hypothesis. Before PHFs are formed in the course of neuronal degeneration, tau must presumably be detached from microtubules. This could be achieved by phosphorylation, for example at Ser-262 by the kinase MARK. Detachment of tau from microtubules would lead to their destabilization and the subsequent breakdown of the microtubule-based axonal transport, which would impair the neuron's viability. The detached tau could, in turn, dimerize by oxidation and aggregate into PHFs. The roles of key residues Ser-262 and Cys-322 are summarized in Fig. 4.

In this context it is interesting to note that oxidation increases in aging tissues (Smith et al. 1991), and that the toxicity of the Aβ peptide is mediated by oxidation (Behl et al. 1994). Furthermore, the observation that fetal tau has only three repeats (and thus only Cys-322) and is phosphorylated in a manner reminiscent of PHF tau (Kanemaru et al. 1992; Bramblett et al. 1993) leads one to speculate that the "fetal" state of tau is particularly prone to nucleate

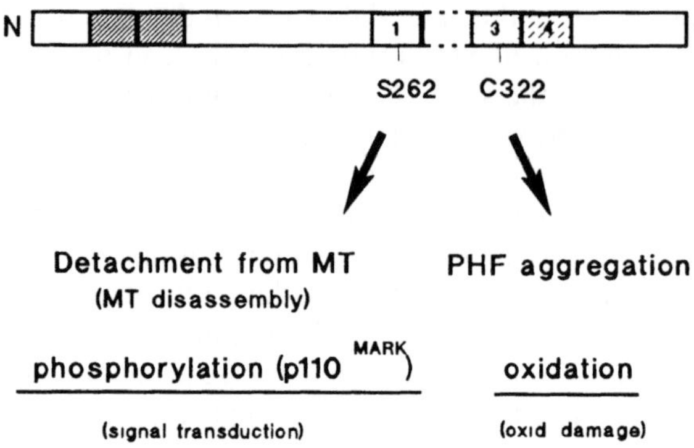

Fig. 4. Two key residues in tau protein and their roles identified so far Ser-262 can be phosphorylated and can regulate the affinity for microtubules (*MT*), this residue can be phosphorylated by the kinase p110 (MARK) Cys-322 can be oxidized and promotes the assembly into paired helical filaments (*PHFs*) Both residues lie within the repeat domain

PHF formation. This could occur as a result of some toxic insult that causes the neuron to defend itself by increased sprouting (Masliah et al. 1991). Conversely, if one wanted to prevent PHF formation, the data suggest two strategies, at least theoretically: 1) by increasing the percentage of four-repeat (adult) tau, because four repeats are inhibitory for the formation of tau dimers and PHFs, even when they become oxidized; and 2) by reducing the oxidative stress, e.g., by antioxidants.

Since the analysis of PHF formation is still in its early stages, it may be worth pointing to some yet unanswered questions.

1. We have postulated a role for the oxidized tau dimer in nucleating PHFs, but the subsequent elongation must involve a different and unknown type of interaction, since no further sulfhydrils are available for crosslinking by oxidation. Assembly via oligomeric stages is a common phenomenon in biological self-assembly systems, and it is possible that useful concepts may emerge by comparing tau's assembly with that of other proteins, for example intermediate filaments (e.g., Heins et al. 1993).
2. The assembly of PHFs in vitro occurs preferentially by three-repeat constructs; however, in vivo PHFs show a similar composition of all six human tau isoforms as the normal tau pool (i.e., both three- and four-repeat forms; Goedert et al. 1992). The incorporation of four-repeat forms will have to be addressed by future experiments.
3. The data described here were based on tau constructs comprising essentially the repeat domain, whereas full-length tau does not form bona fide PHFs efficiently and shows a lower tendency to aggregate (Degarcini et al. 1988; Lichtenberg et al. 1990; Crowther et al. 1994). It is currently not clear how full-length tau can be added to PHFs, but it is conceivable that phosphorylation plays a role at this stage.

4. The structural basis of the tau-tau interaction in PHFs has not yet been established. We know that tau can form antiparallel dimers (Wille et al. 1992), that four-repeat constructs are folded so that the two Cys 291 and 322 are near one another (Schweers et al. 1995), and that neither tau nor PHFs have substantial secondary structure (Schweers et al. 1994). On the other hand, a structure as periodic as PHFs is almost certainly based on some specific subunit interactions. If these were known it would be possible to design strategies against PHF formation in a more rational manner.

Finally, we note an interesting correlation between isoforms of tau and of apolipoprotein E. ApoE4 has been shown to be a major risk factor for AD (Corder et al. 1993). It differs from ApoE3 and ApoE2 in the number of Cys residues (none in E4, 1 or 2 in E3 or E2). It has been proposed that the effect of ApoE is based on differential interactions with tau protein (see Strittmatter et al. 1994, and elsewhere in this volume). This would point to a role in sulfhydril oxidation, and in this sense the reactions of ApoE and tau might both depend on oxidative stress in aging neurons. A change in protein structure, reaction mechanism, or assembly properties could affect the direct interaction between ApoE and tau, as proposed by Strittmatter and colleagues, and/or it could involve other cytoplasmic or extracellular proteins.

Acknowledgments. We would like to thank Dr. M. Goedert for clones of human tau and the Bryan Alzheimer Disease Research Center (Duke University Medical Center, Durham, NC), the Brain Tissue Resource Center (McLean Hospital/Harvard Medical School, Belmont, MA), and the Alzheimer Research Center (Univ. Rochester Medical School, Rochester, NY) for Alzheimer brain tissue samples. This work was supported by Bundesministerium für Forschung und Technologie (BMFT) and the Deutsche Forschungsgemeinschaft (DFG).

References

Aizawa H, Kawasaki H, Murofushi H, Kotani S, Suzuki K, Sakai H (1988) Microtubule-binding domain of tau proteins J Biol Chem 263 7703–7707

Behl C, Davis JB, Lesley R, Schubert D (1994) Hydrogen peroxide mediates amyloid beta protein toxicity Cell 77 817–827

Biernat J, Gustke N, Drewes G, Mandelkow E-M, Mandelkow E (1993) Phosphorylation of serine 262 strongly reduces the binding of tau protein to microtubules Distinction between PHF-like immunoreactivity and microtubule binding Neuron 11: 153–163

Braak E, Braak H, Mandelkow E-M (1994) A sequence of cytoskeleton changes related to the formation of neurofibrillary tangles and neuropil threads Acta Neuropathol 87. 554–567

Bramblett GT, Goedert M, Jakes R, Merrick SE, Trojanowski JQ, Lee VMY (1993) Abnormal tau phosphorylation at Ser(396) in Alzheimer's disease recapitulates development and contributes to reduced microtubule binding Neuron 10 1089–1099

Brandt R, Lee G (1993) Functional organization of microtubule-associated protein tau: identification of regions which affect microtubule growth, nucleation, and bundle formation in vitro. J Biol Chem 268 3414–3419

Butler M, Shelanski ML (1986) Microheterogeneity of microtubule-associated tau-proteins is due to differences in phosphorylation J Neurochem 47 1517–1522

Butner KA, Kirschner MW (1991) Tau-protein binds to microtubules through a flexible array of distributed weak sites J Cell Biol 115 717–730

Corder EH, Saunders AM, Strittmatter WJ, Schmechel DE, Gaskell PC, Small GW, Roses AD, Haines JL, Pericak-Vance MA (1993) Gene dose of apolipoprotein-E type 4 allele and the risk of Alzheimers disease in late-onset families Science 261 921–923

Crowther RA, Olesen OF, Smith MJ, Jakes R, Goedert M (1994) Assembly of Alzheimer-like filaments from full-length tau protein. FEBS Letters 337 135–138

Degarcini EM, Carrascosa JL, Correas I, Nieto A, Avila J (1988) Tau-factor polymers are similar to paired helical filaments of Alzheimer's disease. FEBS Lett 236 150–154

Drewes G, Trinczek B, Illenberger S, Biernat J, Schmitt-Ulms G, Meyer HE, Mandelkow E-M, Mandelkow E (1995) MAP/microtubule affinity regulating kinase P110(mark) a novel protein kinase that regulates tau-microtubule interactions and dynamic instability by phosphorylation at the Alzheimer-specific site Serine 262 J Biol Chem 270· 7679–7688

Ennulat DJ, Liem RKH, Hashim GA, Shelanski ML (1989) Two separate 18-amino acid domains of tau promote the polymerization of tubulin J Biol Chem 264 5327–5330

Glenner GG, Eanes ED, Page DL (1972) The relation of the properties of Congo red-stained amyloid fibrils to the beta-conformation J Histochem Cytochem 20 821–826

Goedert M, Spillantini M, Jakes R, Rutherford D, Crowther RA (1989) Multiple isoforms of human microtubule-associated protein-tau sequences and localization in neurofibrillary tangles of Alzheimers-disease Neuron 3· 519–526

Goedert M, Spillantini G, Cairns NJ, Crowther RA (1992) Tau proteins of Alzheimer paired helical filaments abnormal phosphorylation of all six brain isoforms Neuron 8: 159–168

Grundke-Iqbal I, Iqbal K, Tung Y, Quinlan M, Wisniewski H, Binder L (1986) Abnormal phosphorylation of the microtubule-associated protein tau in Alzheimer cytoskeletal pathology Proc Natl Acad Sci USA 83 4913–4917

Gustke N, Trinczek B, Biernat J, Mandelkow E-M, Mandelkow E (1994) Domains of tau protein and interactions with microtubules. Biochemistry 33· 9511–9522

Hasegawa M, Morishima-Kawashima M, Takio K, Suzuki M, Titani K, Ihara Y (1992) Protein sequence and mass spectrometric analyses of tau in the Alzheimer's disease brain. J Biol Chem 267. 17047–17054

Heins S, Wong PC, Muller S, Goldie K, Cleveland DW, Aebi U (1993) The rod domain of NF-L determines neurofilament architecture, whereas the end domains specify filament assembly and network formation J Cell Biol 123 1517–1533

Inouye H, Fraser PE, Kirschner DA (1993) Structure of beta-crystallite assemblies formed by Alzheimer beta-amyloid protein analogues. analysis by x-ray diffraction Biophys J 64 502–519

Kanemaru K, Takio K, Miura R, Titani K, Ihara Y (1992) Fetal-type phosphorylation of the tau in paired helical filaments J Neurochem 58: 1667–1675

Kondo J, Honda T, Mori H, Hamada Y, Miura R, Ogawara M, Ihara Y (1988) The carboxyl third of tau is tightly bound to paired helical filaments. Neuron 1: 827–834

Kosik KS, McConlogue L (1994) Microtubule-associated protein function lessons from expression in Spodoptera frugiperda cells Cell Mot Cytoskel 28· 195–198

Lichtenberg-Kraag B, Mandelkow E-M (1990) Isoforms of tau protein from mammalian brain and avian erythrocytes structure, self-assembly, and elasticity J Struct Biol 105: 46–53

Lichtenberg-Kraag B, Mandelkow E-M, Biernat J, Steiner B, Schroter C, Gustke N, Meyer HE, Mandelkow E (1992) Phosphorylation dependent interaction of neurofilament antibodies with tau protein. Epitopes, phosphorylation sites, and relationship with Alzheimer tau Proc Natl Acad Sci USA 89: 5384–5388

Mandelkow E-M, Mandelkow E (1993) Tau as a marker for Alzheimer's disease TIBS 18· 480–483

Mandelkow E-M, Biernat J, Lichtenberg-Kraag B, Drewes G, Wille H, Gustke N, Baumann K, Mandelkow E (1995a) Phosphorylation of tau and relationship with Alzheimer paired helical filaments In Kosik K, Christen Y, Selkoe D (eds) Alzheimer's disease. lessons from cell biology Springer, Berlin Heidelberg New York Tokyo, pp 103–120

Mandelkow E, Song Y-H, Mandelkow E-M (1995b) The microtubule lattic – dynamic instability of concepts. Trends in Cell Biol 5· 262–266

Masliah E, Mallory M, Hansen L, Alford M, Albright T, Deteresa R, Terry R, Baudier J, Saitoh T (1991) Patterns of aberrant sprouting in Alzheimer's disease Neuron 6 729–739

Morishima-Kawashima M, Hasegawa M, Takio K, Suzuki M, Titani K, Ihara Y (1993) Ubiquitin is conjugated with amino-terminally processed tau in paired helical filaments Neuron 10· 1151–1160

Schweers O, Schonbrunn E, Marx A, Mandelkow E (1994) Structural studies of tau protein and Alzheimer paired helical filaments show no evidence for beta structure J Biol Chem 269 24290–24297

Schweers O, Mandelkow E-M, Biernat J, Mandelkow E (1995) Oxidation of cysteine 322 in the repeat domain of microtubule-associated protein tau controls the assembly of Alzheimer paired helical filaments Proc Natl Acad Sci USA, in press

Smith CD, Carney JM, Starke-Reed PE, Oliver CN, Stadtman ER, Floyd RA, Markesbery WR (1991) Excess brain protein oxidation and enzyme dysfunction in normal aging and in Alzheimer's disease Proc Natl Acad Sci USA 88 10540–10543

Strittmatter WJ, Saunders AM, Goedert M, Weisgraber KH, Dong LM, Jakes R, Huang DY, Pericak-Vance M, Schmechel D, Roses AD (1994) Isoform-specific interactions of apolipoprotein-E with microtubule-associated protein tau· implications for Alzheimer disease Proc Natl Acad Sci USA 91 11183–11186

Wille H, Drewes G, Biernat J, Mandelkow E-M, Mandelkow E (1992) Alzheimer-like paired helical filaments and antiparallel dimers formed from microtubule-associated protein tau in vitro J Cell Biol 118 573–584

Wischik C, Crowther R, Stewart M, Roth M (1985) Subunit structure of paired helical filaments in Alzheimer's disease J Cell Biol 100· 1905–1912

Wischik C, Novak M, Thogersen H, Edwards P, Runswick M, Jakes R, Walker J, Milstein C, Roth M, Klug A (1988) Isolation of a fragment of tau derived from the core of the paired helical filament of Alzheimer disease Proc Natl Acad Sci USA 85. 4506–4510

Yan SD, Chen X, Schmidt A, Brett J, Godman G, Zou Y, Scott CW, Caputo C, Frappier T, Smith MA, Perry G, Yen SH (1994) Glycated tau protein in Alzheimer-disease a mechanism for induction of oxidant stress Proc Natl Acad Sci USA 91 7787–7791

The βA4 Amyloid Peptide Enhances Cellular Uptake of Apolipoprotein E-Containing Lipoproteins by the Low-Density Lipoprotein Receptor-Related Protein/α_2-Macroglobulin Receptor

H. Scharnagl, K. Winkler, M. Hüttinger, D. Wittmann, M. Nauck, H. Wieland, W. Groß, T.G. Ohm,* and *W. März*

Summary

We incubated β-very low density lipoprotein (β-VLDL) from cholesterol-fed rabbits with βA4 amyloid peptide (43mer). βA4 associated with the apolipoprotein E moiety of the βVLDL. The binding of βA4 to apolipoprotein E could not be dissociated with sodium dodecyl sulphate. Unlike native βVLDL, the βA4/βVLDL complexes bound to the low-density lipoprotein (LDL) receptor-related protein of normal and familial hypercholesterolemia fibroblasts in culture. The βA/βVLDL complexes were taken up into the cells, but escaped proteolytic degradation. The cellular uptake of βA4/βVLDL complexes was independent from the LDL receptor pathway; it was inhibited by an excess of unlabelled, activated α_2-macroglobulin. βA4 competed with the binding and the uptake, but not the degradation, of ^{125}I-labelled α_2-macroglobulin in fibroblasts. Considered together, the data suggest that the association of βA4 with apolipoprotein E-containing lipoproteins gives rise to complexes that are endocytosed through the LDL receptor-related protein but fails to undergo proteolytic degradation in the lysosomes. These findings may be relevant to the pathobiochemistry of Alzheimer's disease.

Introduction

The low-density lipoprotein receptor-related protein (LRP) is a multifunctional cell surface receptor (Krieger and Herz 1994). It consists of 4544 amino acids (Herz et al. 1988) and is expressed on the surface of many cultured cells and in many organs (Herz et al. 1988). LRP has been implicated in the clearance from plasma of apolipoprotein (apo) E-containing lipoproteins (Krieger and Herz 1994). Kristensen et al. (1990) and Strickland et al. (1990) reported that LRP was identical to the receptor for activated α_2-macroglobulin (α_2M). LRP serves as a receptor for a number of further ligands such as tissue plasminogen activator, urokinase plasminogen activator-inhibitor complexes, tissue plasminogen activator-inhibitor complexes, lactoferrin, lipoprotein

* Division of Clinical Chemistry, Department of Medicine, Albert-Ludwigs University, Hugstetter Strasse 55, 70106 Freiburg, Germany

A D Roses et al (Eds)
Apolipoprotein E and Alzheimer's Disease

lipase, lipoprotein (a), and *Pseudomonas aeruginosa* exotoxin A (Krieger and Herz 1994; März et al. 1993a).

The binding of lipoproteins to LRP is mediated by apoE (Mahley 1988). ApoE is a 34-kDa glycoprotein that is associated with triglyceride-rich lipoproteins and high density lipoproteins (Mahley 1988). By virtue of its ability to bind to the low density lipoprotein (LDL) receptor and to LRP, it plays a key role in lipoprotein metabolism. ApoE is polymorphic in sequence. The products of the three common alleles (ε2, ε3, ε4) differ from one another at positions 112 and 158 of the amino acid sequence. ApoE3, the most frequent isoform, has arginine at position 112 and cysteine at position 158. ApoE4 has arginine and apoE2 has cysteine at both positions (Weisgraber et al. 1981). The ε4 allele is associated with elevated LDL cholesterol and an increased risk of atherosclerosis; the ε2 allele, in contrast, lowers LDL cholesterol and cardiovascular risk (Utermann 1987; Davignon et al. 1988).

In the central nervous system, apoE is produced and secreted by astrocytes; in the peripheral nervous system nonmyelinating glial cells express apoE (Boyles et al. 1985). Following nerve injury, the synthesis of apoE is dramatically increased, suggesting a role in the provision of lipids during repair processes (Boyles et al. 1989; Ignatius et al. 1986). Namba et al. (1991) observed that apoE was deposited in Alzheimer's disease (AD), an irreversible neurodegenerative disorder resulting in progressive dementia. Strittmatter et al. (1993) recently linked the apoE polymorphism to AD. Since then, convincing evidence has accumulated that the ε4 allele greatly enhances the susceptibility for AD (Poirier et al. 1993; Saunders et al. 1993; Rebeck et al. 1993). Along with the formation of neurofibrillary tangles and neuropil threads, the deposition in neuritic plaques of a 39- to 43-residue amyloid beta peptide (βA4) is the major histopathological feature of AD. Autopsy-based studies consistently showed that the A4/β-amyloid load was increased in brains from ε4 carriers (Rebeck et al. 1993; Schmechel et al. 1993), and we found that a positive correlation also exists between the stage of intraneural neurofibrillary changes and the ε4 gene dose (Ohm et al. 1995). According to Strittmatter et al. (1993a,b), βA4 binds to apoE4 with greater affinity than to apoE3, a finding that may in part account for the close association between the apoE polymorphism and AD. Apart from βA4 and apoE, LRP and its ligand α_2M have been localized to senile plaques and reactive astrocytes in AD (Rebeck et al. 1993; Bauer et al. 1991). These observations prompted us to examine whether βA4 was able to modify the interaction of apoE-containing lipoproteins with cultured cells. Using βVLDL from cholesterol-fed rabbits as a model, we showed that βA4 is able to mediate the endocytosis of lipoprotein particles by LRP.

Methods

βVLDL were isolated from the plasma of 1% cholesterol-fed rabbits by preparative ultracentrifugation as described (Kovanen et al. 1981) and were iodinated using *n*-bromosuccinimide as oxidizing agent (Sinn et al. 1988). To generate βA4/βVLDL complexes, we incubated βA4 (43-mer) with βVLDL in phosphate buffered saline at 37 °C, usually for 14 hours. α_2-Macroglobulin (α_2M, Sigma) was activated with methylamine according to Ashcom et al. (1990). Fibroblasts were from skin biopsies of normolipidemic healthy individuals and from a 14-year-old female patient with homozygous familial hypercholesterolemia (FH). Sodium dodecyl sulphate (SDS, final concentration 20 g/l) polyacrylamide gel (T = 10%; C = 5%) electrophoresis was carried out in the absence of disulfide reducing agents after heating the samples for 5 minutes at 100°C using a discontinuous buffer system (Laemmli 1970). Binding, uptake (binding *plus* internalization) and degradation of ^{125}I-labelled βVLDL were determined as described (Goldstein et al. 1983) with slight modifications (März et al. 1993b). The cells were incubated for 1 h at 4 °C to determine surface binding and for 4 h at 37°C to determine uptake and lysosomal degradation.

Results

When we incubated βA4 with βVLDL from cholesterol-fed rabbits for 14 h at 37°C, βA4 complexed to βVLDL. As demonstrated by immunoblotting, βA4 almost exclusively interacted with apoE in βVLDL (data not shown). We incubated βA4 at increasing concentrations with ^{125}I-labelled βVLDL and examined the interaction of the resulting complexes with normal human skin fibroblasts (Fig. 1). When βA4 and ^{125}I-βVLDL were used at final protein

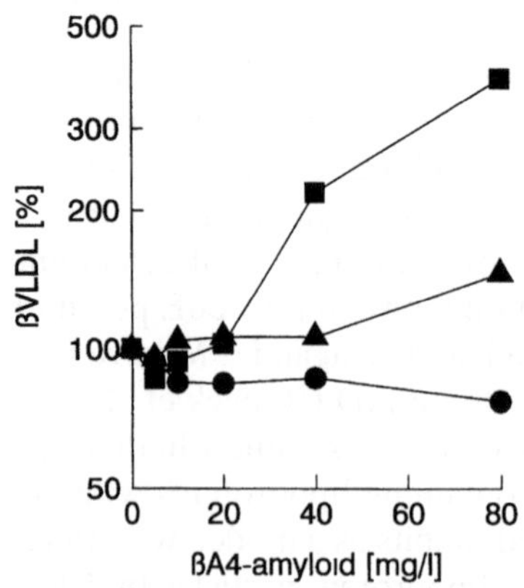

Fig. 1. Binding, uptake, and degradation of β-very low density lipoprotein (*βVLDL*) and βA4/βVLDL complexes in normal human skin fibroblasts ^{125}I-βVLDL at a concentration of 1 mg/l protein was pre-incubated with βA4 at the concentrations indicated on the *abscissa* Normal human skin fibroblasts were grown in RPMI 1640 medium supplemented with 10% (vol/vol) fetal calf serum The cells were incubated with ^{125}I-labelled βVLDL or βA4/βVLDL complexes Cell surface binding at 4°C (*triangles*), uptake (*squares*), and degradation (*circles*) at 37°C were determined as described in "Methods" Each data point represents the average from triplicates

concentrations of 40 mg/l and 1 mg/l, respectively, cell surface binding of βVLDL at 4°C was 46% higher and lysosomal degradation at 37°C was 23% lower, compared to incubations without βA4. The most marked effect was seen on the amount of cell-associated radioactivity, which increased more than twofold in the presence of βA4. The increase in uptake was a function of the time βA4 was allowed to equilibrate with βVLDL. When βA4 and βVLDL were added to the cells without pre-incubation (zero time point), uptake was essentially unaffected. Pre-incubation of βA4 and βVLDL for 30 min however, was sufficient to cause a significant increase in cell-associated radioactivity; an additional increase was detected after 14 h of pre-incubation (Fig. 2). Again, βA4 did not influence the lysosomal degradation of βVLDL, regardless of the pre-incubation time. In LDL receptor-negative fibroblasts, complexation of βA4 to βVLDL enhanced the cellular uptake in a fashion similar to normal fibroblasts, indicating that the LDL receptor was not significantly involved in the endocytosis of βA4/βVLDL complexes (not shown).

We reasoned that, upon complexation to βVLDL, βA4 enhanced the binding and the uptake of βVLDL into the cells. This assumption would be strongly supported if βA4 was itself a ligand for a βVLDL binding site on the surface of the cell. To examine this possibility, we studied whether unlabelled, "free" βA4, i.e., βA4 not previously complexed to βVLDL, competed for the binding and uptake of βVLDL in familial hypercholesterolemia fibroblasts. As shown in Fig. 3, "free" βA4 inhibited the binding of βVLDL at 4°C but

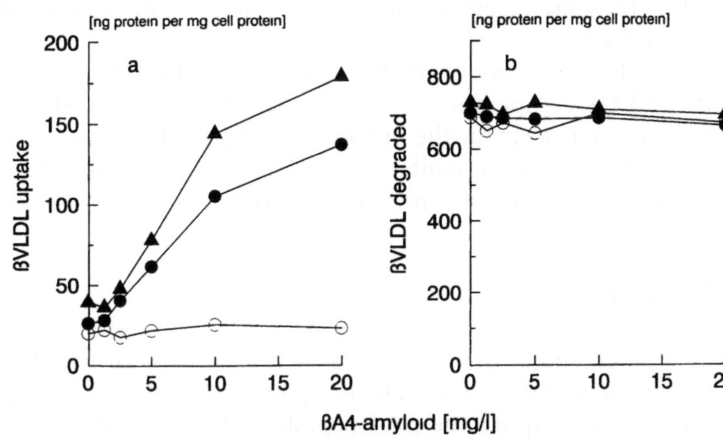

Fig. 2. Effect of the pre-incubation time of βA4 and β-very low density lipoprotein (*βVLDL*) on the uptake of βA4/βVLDL complexes in normal human skin fibroblasts. ^{125}I-βVLDL (0.5 mg/l βVLDL-protein) were pre-incubated for 30 min (*solid circles*) or 14 h (*solid triangles*) with βA4 at the concentrations indicated on the *abscissa*. Normal human skin fibroblasts were grown in RPMI 1640 medium supplemented with 10% (vol/vol) fetal calf serum and incubated with ^{125}I-labelled βVLDL or βA4/βVLDL complexes. Cellular uptake (*left*) and degradation (*right*) were measured at 37°C as described in "Methods". Control incubations were performed in which βVLDL and βA4 were added to the cells without pre-incubation (*open circles*) Each data point represents the average from triplicates

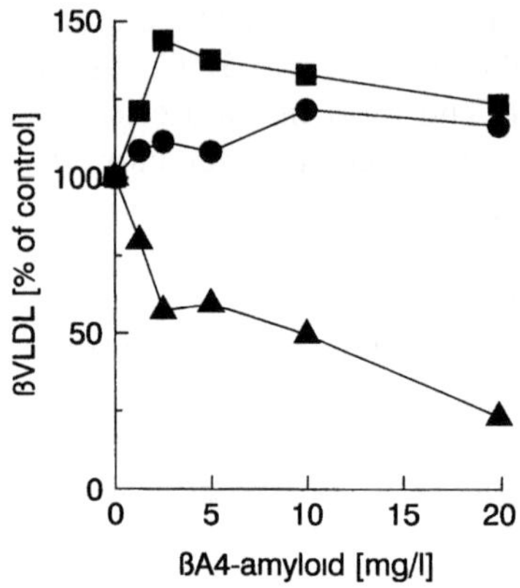

Fig. 3. Competition for ^{125}I-labelled β-very low density lipoprotein ($\beta VLDL$) binding, uptake, and degradation by "free" βA4-amyloid Familial hypercholesterolemia fibroblasts were grown in RPMI 1640 medium supplemented with 10% (vol/vol) fetal calf serum The cells received ^{125}I-labelled βVLDL at a final concentration of 0 3 mg/l and βA4 at the concentrations indicated on the *abscissa* without pre-incubating βVLDL and βA4 Binding at 4°C (*triangles*) and uptake (*squares*) and degradation (*circles*) at 37°C were determined as described in "Methods" Each data point represents the average from triplicates

inhibited neither uptake nor degradation at 37°C. We wished to identify the receptor responsible for the uptake of βA4/βVLDL complexes more directly. LRP is the only receptor known so far that is able to bind and internalize activated α_2M.

We therefore tested whether α_2M was able to compete for the uptake of βA4/βVLDL complexes. As shown in Fig. 4, activated α_2M was highly effective as a competitor for the uptake of βA4/βVLDL complexes. Consistently, βA4 alone was able to inhibit the binding at 4°C and the uptake, but not the degradation at 37°C, of activated ^{125}I-α_2M.

Discussion

In the circulation, apoE functions as a ligand for at least two cell surface lipoprotein receptors, the LDL receptor and LRP (Krieger and Herz 1994; Mahley 1988). ApoE, the βA4 peptide, LRP and α_2M co-localize in senile plaques of brains affected with AD (Rebeck et al. 1993; Schmechel et al. 1993; Bauer et al. 1991). Therefore, we investigated the influence of βA4 on the cellular uptake and processing of apoE-containing lipoproteins.

Our data demonstrate that βA4, when allowed to equilibrate with βVLDL, enhances the cellular uptake of βVLDL in a concentration-dependent fashion. The effect of βA4 on βVLDL uptake positively correlated with the time βA4 was pre-incubated with βVLDL at 37°C. Previous studies have shown that native βVLDL are not endocytosed by LRP unless being enriched

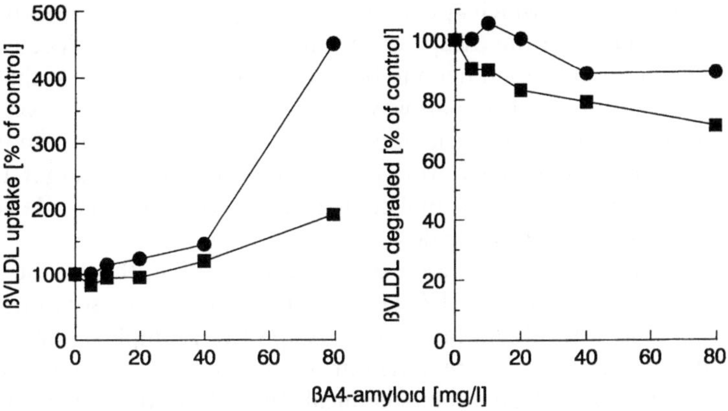

Fig. 4. Effect of unlabelled activated α_2-macroglobulin (α2M) on the uptake and degradation of βA4/-β-very low density lipoprotein ($\beta VLDL$) complexes in normal human skin fibroblasts Cells were grown in RPMI 1640 medium supplemented with 10% (vol/vol) fetal calf serum βA4/βVLDL adducts were generated by incubating ^{125}I-βVLDL at a final concentration of 1 mg/l protein with βA4 at the concentration indicated on the *abscissa* for 14 h Uptake (*left*) and degradation (*right*) of βA4/βVLDL complexes were determined in the absence (*circles*) and in the presence (*squares*) of activated α_2M (50 mg/l), each data point represents the average from triplicates

in vitro with apoE (Kowal et al. 1989; Hussain et al. 1991) or lipoprotein lipase (Willnow et al. 1992). The present data provide evidence that complexation of βA4 to βVLDL generates an adduct with high avidity towards LRP. βA4 is able to effectively compete with activated α_2M for cellular uptake, indicating that βA4 itself represents a ligand of LRP. Using unlabelled α_2M as a competitor for βA4/^{125}I-βVLDL complexes, we confirmed that the βA4-mediated rise in uptake of βVLDL was almost completely accounted for by LRP.

We wish to emphasize that complexes of βVLDL and βA4 displayed greatly enhanced endocytosis but obviously did not undergo degradation. Current opinion predicates that LRP, similar to the LDL receptor, delivers internalized apoE-containing lipoproteins to an acidic compartment such as the endosome, where the receptor molecule undergos a pH-dependent conformational transition that results in the release of the ligand (Herz et al. 1990). The receptor is then recycled to the cell surface and the ligand is forwarded to the lysosomes for degradation (Kowal et al. 1989; Hussain et al. 1991). Our results show that complexation of βA4 to apoE-containing βVLDL stimulates cellular uptake severalfold, but has no effect on the degradation. This reflects that the βA4/apoE complexes fail to reach the lysosomal compartment; it may be significant for the role of apoE in the nervous system and for the pathobiochemistry of AD. It has been postulated that apoE normally binds to tau protein, thereby allowing it to stabilize microtubules (Strittmatter et al. 1994). In the brain, apoE mRNA is found in astrocytes, but not in neurons. As neurons contain immunoreactive apoE (Han et al. 1994), a metabolic pathway has to exist by which apoE is taken up. Selkoe (1994) demanded that "a novel mechanism for the escape of the endocytosed apoE molecule from lysosomal degradation" would need to be found. Our results suggest that LRP is part of such an endocytotic pathway and that the βA4 peptide, once bound to an apoE-containing particle, mimics the function of lipoprotein lipase in triggering endocytosis. However, in that βA4 appears to target the internalized particle to a non-lysosomal compartment, its regulatory role is clearly distinct from the former molecule.

Our findings raise a series of questions. First, the precise molecular characteristics of the βA4/βVLDL complexes need to be scrutinized further. Second, the intracellular fate of the βA4/βVLDL complexes has to be analyzed in detail. Third, studies relating our observations to different apoE isoforms are required. Finally, pathobiochemical implications regarding neurodegenerative diseases will have to be worked out.

Acknowledgments. We thank Dr. P. Grützmacher (Department of Internal Medicine II, St. Markus-Hospital, Frankfurt am Main) for the skin biopsy from the patient with homozygous familial hypercholesterolemia and Bettina Donnerhak, Brigitte Haas, Sybille Rall and Ulrike Stein for technical assistance; the financial support from Boehringer Ingelheim to M.H. and W.M. is gratefully acknowledged.

References

Ashcom JD, Tiller SE, Dickerson K, Cravens JL, Argraves WS, Strickland DK (1990) The human α_2-macroglobulin receptor Identification of a 420-kD cell surface glycoprotein specific for the activated conformation of α_2 macroglobulin J Cell Biol 110 1041–1048

Bauer J, Strauss S, Schreiter-Gasser U, Ganter U, Schlegel P, Witt I, Volk B, Berger M (1991) Interleukin-6 and α_2-macroglobulin indicate an acute-phase state in Alzheimer's disease cortices FEBS Lett 285 111–114

Boyles JK, Pitas RE, Wilson E, Mahley RW, Taylor JM (1985) Apolipoprotein E associated with astrocytic glia of the central nervous system and with nonmyelinating glia of the peripheral nervous system J Clin Invest 76 1501–1513

Boyles JK, Zoellner CD, Anderson LJ, Kosik LM, Pitas RE, Weisgraber KH, Hui DY, Mahley RW, Gebicke-Haerter PJ, Ignatius MJ, Shooter EM (1989) A role for apolipoprotein E, apolipoprotein A-I, and low density lipoprotein receptors in cholesterol transport J Clin Invest 83: 1015–1031

Davignon J, Gregg RE, Sing CF (1988) Apolipoprotein E polymorphism and atherosclerosis Arteriosclerosis 8 1–21

Goldstein JL, Basu SK, Brown MS (1983) Receptor-mediated endocytosis of low-density lipoprotein in cultured cells In Fleischer S, Fleischer B (eds) Biomembranes Part L Membrane biogenesis. Methods in enzymology, Vol 98 Academic Press, Orlando, pp 241–260

Han SH, Einstein G, Weisgraber KH, Strittmatter WJ, Saunders A, Pericak-Vance M, Roses AD, Schmechel DE (1994) Apolipoprotein E is localized to the cytoplasm of human cortical neurons a light and electron microscopic study J Neuropathol Exp Neurol 53 535–544

Herz J, Hamann U, Rogne S, Myklebost O, Gausepohl H, Stanley KK (1988) Surface location and high affinity for calcium of a 500 kD liver membrane protein closely related to the LDL-receptor suggest a physiological role as lipoprotein receptor EMBO J 7 4119–4127

Herz J, Kowal RC, Ho JK, Brown MS, Goldstein JL (1990) Low density lipoprotein receptor-related protein mediates endocytosis of monoclonal antibodies in cultured cells and in rabbit liver J Biol Chem 265 21355–21362

Hussain MM, Maxfield FR, Mas-Oliva J, Tabas I, Ji Z-S, Innerarity TL, Mahley RW (1991) Clearance of chylomicron remnants by the low density lipoprotein receptor-related protein/α_2-macroglobulin receptor. J Biol Chem 226 13936–13940

Ignatius MJ, Gebicke-Haerter PJ, Skene JHP, Schilling JW, Weisgraber KH, Mahley RW, Shooter EM (1986) Expression of apolipoprotein E during nerve degeneration and regeneration Proc Natl Acad Sci USA 83 1125–1129

Kovanen PT, Brown MS, Basu SK, Bilheimer DW, Goldstein JL (1981) Saturation and suppression of hepatic lipoprotein receptors a mechanism for the hypercholesterolemia of cholesterol-fed rabbits Proc Natl Acad Sci USA 78 1396–1400

Kowal RC, Herz J, Goldstein JL, Esser V, Brown MS (1989) Low density lipoprotein receptor-related protein mediates uptake of cholesteryl esters derived from apoprotein E-enriched lipoproteins Proc Natl Acad Sci USA 86 5810–5814

Krieger M, Herz J (1994) Structures and functions of multiligand lipoprotein receptors. macrophage scavenger receptors and LDL receptor related-protein (LRP) Annu Rev Biochem 63 601–637

Kristensen T, Moestrup SK, Glieman J, Bendtsen L, Sand O, Sottrup-Jensen L (1990) Evidence that the newly cloned low-density-lipoprotein receptor related protein (LRP) is the α_2-macroglobulin receptor FEBS Lett 276 151–155

Laemmli UK (1970) Cleavage of structural proteins during assembly of the head of bacteriophage T4 Nature 227 680–685

Mahley RW (1988) Apolipoprotein E Cholesterol transport protein with expanding role in cell biology Science 240 622–630

Marz W, Beckmann A, Scharnagl H, Siekmeier R, Mondorf U, Held I, Schneider W, Preissner KT, Curtiss LK, Groß W, Huttinger M (1993a) Heterogeneous lipoprotein (a) size isoforms differ by their interaction with the low density lipoprotein receptor and the low density lipoprotein receptor-related protein/L_2-macroglobulin FEBS Lett 325 271–275

Marz W, Baumstark MW, Scharnagl H, Ruzicka V, Buxbaum S, Herwig J, Pohl T, Russ A, Schaaf L, Berg A, Bohles H-J, Usadel KH, Groß W (1993b) Accumulation of "small dense" low density lipoproteins in a homozygous patient with familial defective apolipoprotein B-100 results from heterogeneous interaction of LDL-subfractions with the LDL receptor J Clin Invest 92. 2922–2933

Namba Y, Tomonaga M, Kawasaki H, Otomo EKI (1991) Apolipoprotein E immunoreactivity in cerebral amyloid deposits and neurofibrillary tangles in Alzheimer's disease and kuru plaque amyloid in Creutzfeld-Jakob disease Brain Res 541 163–166

Ohm TG, Kirca M, Bohl J, Scharnagl H, Groß W, Marz W (1995) ApoE influences not only cerebral senile plaque load, but also Alzheimer-type neurofibrillary tangles Neuroscience 66 583–587

Poirier J, Davignon J, Bouthillier D, Kogan S, Bertrand P, Gauthier S (1993) Apolipoprotein E polymorphism and Alzheimer's disease Lancet 342· 697–699

Rebeck GW, Reiter JS, Strickland DK, Hyman BT (1993) Apoliporotein E in sporadic Alzheimer's disease: Allelic variation and receptor interactions Neuron 11: 575–580

Saunders AM, Strittmatter WJ, Schmechel D, St George-Hyslop PH, Pericak-Vance MA, Joo SH, Rosi BL, Gusella JF, Crapper-MacLachlan DR, Alberts MJ, Hulette C, Crain B, Goldgaber D, Roses AD (1993) Association of apolipoprotein E allele ⊢4 with late-onset familial and sporadic Alzheimer's disease. Neurology 43 1467–1472

Schmechel DE, Saunders AM, Strittmatter WJ, Crain BJ, Hulette CM, Joo SH, Pericak-Vance MA, Goldgaber D, Roses AD (1993) Increased amyloid β-peptide deposition in cerebral cortex as a consequence of apolipoprotein E genotype in late-onset Alzheimer disease Proc Natl Acad Sci USA 90 9649–9653

Selkoe DJ (1994) Alzheimer's disease: a central role for amyloid J Neuropathol Exp Neurol 53 438–447

Sinn HJ, Schrenk HH, Friedrich EA, Via DP, Dresel HA (1988) Radioiodination of proteins and lipoproteins using n-bromosuccinimide as oxidizing reagent Anal Biochem 170 186–192

Strickland DK, Ashcom JD, Williams S, Burgess WH, Migliorini M, Argraves WS (1990) Sequence identity between the α_2-macroglobulin receptor and low density lipoprotein receptor-related protein suggests that this molecule is a multifunctional receptor J Biol Chem 265 17401–17404

Strittmatter WJ, Saunders AM, Schmechel D, Pericak-Vance M, Enghild J, Salvesan GS, Roses AD (1993a) Apolipoprotein E high-avidity binding to β-amyloid and increased frequency of type 4 allele in late-onset familial Alzheimer disease Proc Natl Acad Sci USA 90 1977–1981

Strittmatter WJ, Weisgraber KH, Huang DY, Dong L-M, Salvesen GS, Pericak-Vance M, Schmechel D, Saunders AM, Goldgaber D, Roses AD (1993b) Binding of human apolipoprotein E to synthetic amyloid β peptide isoform-specific effects and implications for late-onset Alzheimer disease Proc Natl Acad Sci USA 90 8098–8102

Strittmatter WJ, Weisgraber KH, Goedert M, Saunders AM, Huang D, Corder EH, Dong L-M, Jakes R, Alberts MJ, Gilbert JR, Schmechel DE, Pericak-Vance MA, Roses AD (1994) Hypothesis microtubule instability and paired helical filament formation in Alzheimer disease brain as a function of apolipoprotein E genotype Exp Neurol 125 163–171

Utermann G (1987) Apolipoprotein E polymorphism in health and disease. Am Heart J 113 433–440

Weisgraber KH, Rall SC, Mahley RW (1981) Human E apoprotein heterogeneity Cysteine arginine interchanges in the amino acid sequence of the apoE isoforms J Biol Chem 256 9077–9083

Willnow TE, Goldstein JL, Orth K, Brown MS, Herz J (1992) Low density lipoprotein receptor-related protein and gp330 bind similar ligands, including plasminogen activator-inhibitor complexes and lactoferrin, an inhibitor of chylomicron remnant clearance J Biol Chem 267 26172–26180

Interactions of Apolipoprotein E with Microtubule-Associated Proteins

W.J. Strittmatter, V. de Serrano, D.Y. Huang, M.E. Martin, A.D. Roses,* and *A.M. Saunders*

Summary

The type 4 allele of apolipoprotein E (APOE, gene; apoE, protein) has been identified as a genetic risk factor for late-onset sporadic and familial Alzheimer's disease (AD). APOE4 raises the risk for AD and lowers the mean age of onset in a gene dose-dependent manner, whereas APOE2, a relatively rare allele, decreases the risk and increases the age of onset of disease. The role of apoE in the pathogenesis of AD is not yet known. We have proposed that the interaction of apoE3 with tau may regulate the function and metabolism of tau by binding and sequestering tau from forming paired helical filaments or from phosphorylation. To test this hypothesis, we are studying the interaction between apoE isoforms and tau using standard in vitro assays followed by western blotting and the relatively new technique of real-time bimolecular interaction analysis.

Introduction

The identification of APOE4 as a genetic susceptibility gene for late-onset familial and sporadic AD (Strittmatter et al. 1993; Saunders et al. 1993; Corder et al. 1993) has led to pathological studies of the distribution of apoE immunoreactivity in brain and to in vitro studies of the interaction of apoE isoforms with AD-related proteins. Our laboratory has been particulary interested in the binding of apoE isoforms to tau, the primary component of the paired helical filaments that are a characteristic of AD. Initial experiments demonstrated that apoE3, the most common isoform, forms an irreversible complex with tau, whereas apoE4 does not. We are currently focusing on the binding of apoE isoforms and tau using an innovative technique known as real-time bimolecular interaction analysis. The data generated from these experiments will help test our hypothesis that the binding of apoE3 to tau may slow the rate

*Departments of Medicine (Neurology) and Neurobiology, Joseph and Kathleen Bryan Alzheimer's Disease Research Center, P.O Box 2900, Duke University Medical Center, Durham, NC 27710, USA

A D Roses et al (Eds)
Apolipoprotein E and Alzheimer's Disease

of tau self-assembly into paired helical filaments (PHFs) and may affect tau phosphorylation.

Biochemical Overview

Pathological studies of apoE immunoreactivity in AD-affected and control brains have identified apoE in plaque and vascular amyloid and within neurons (Diedrich et al. 1991; Namba et al. 1991; Wisniewski and Frangione 1992; Strittmatter et al. 1993; Schmechel et al. 1993; Han et al. 1994a). A minority of the apoE-positive neurons contain neurofibrillary tangles (NFTs). NFTs are abnormal intracellular structures composed of PHFs which, in turn, are made of self-assembled, hyperphosphorylated tau (Lee et al. 1991; Goedert et al. 1992; Goedert 1993). Tau can be lightly phosphorylated in normal brain, and its primary function is microtubule stabilization through binding to tubulin subunits. The abnormally hyperphosphorylated tau that forms PHFs is unable to bind to tubulin; dephosphorylation of PHF-tau will enable it to bind (Bramblett et al. 1993; Yoshida and Ihara 1993).

Under normal circumstances, apoE in the brain is synthesized by astrocytes and microglia and may enter neurons by binding to the low density lipoprotein (LDL) receptor-related protein (LRP) located at the synapse, or other LDL and very low density lipoprotein (VLDL) receptor mechanisms (Kounnas et al. 1992; Weisgraber 1994; Hyman et al., in this volume; Ikeda et al., this volume; Poirier et al., this volume). Alternatively, apoE may be produced within certain neurons, although there is currently no evidence to support this possibility. Electron microscopic analyses of human brain have demonstrated that a fraction of the small intraneuronal pool of apoE is located within the cytoplasmic compartment (Han et al. 1994b). Since tau and apoE have been identified within neuronal cytoplasm, we have proposed that isoform-specific interactions of apoE with tau may in part determine the rates of neuronal injury or repair, determining the age of onset of AD (Strittmatter et al. 1994a).

Binding studies have shown an isoform-specific difference in the in vitro binding of human apoE to recombinant tau and hyperphosphoylated tau (Strittmatter et al. 1994b). Tau binds to apoE3 over a wide range of pH values and concentrations, but does not bind to apoE4. The apoE3/tau complex is sodium dodecyl sulfate (SDS) stable, but can be dissociated by treatment with the reducing agent β-mercaptoethanol. Neither apoE3 nor apoE4 binds to tau that has been hyperphosphorylated in vitro by incubation with a crude rat brain extract. Microtubule-associated protein (MAP2c), which is localized to dendrites, also demonstrates similar isoform-specific binding to apoE3 (Huang et al. 1994).

Similar studies using fragments of apoE and tau have demonstrated the regions of each molecule that are critical for complex formation (Strittmatter et al. 1994b). The tau binding region of apoE3 corresponds to its LDL-receptor binding domain. This domain is distinct from the region of apoE that binds lipoprotein particles and amyloid Aβ peptide. ApoE3 binds to the microtubule-binding repeat region of tau, the same region that appears to promote self-assembly of tau into PHFs. The tau microtubule-binding region is composed of four domains. Each domain consists of either 31 or 32 amino acids; the first 18 residues bind to tubulin. Binding experiments using synthetic tau domain peptides and apoE demonstrated that each of the four domains interacts with apoE3 over a wide range of concentrations, whereas only domain III complexes with apoE4 at the highest concentrations tested (Huang et al. 1995). Further insights into the tau-apoE interaction were developed with these data. The sensitivity of the apoE3/tau complex to β-mercaptoethanol suggested that the single cycteine residue in apoE3 was involved. However, neither domain I nor IV has a cysteine residue, indicating that the binding of apoE3 to tau is not disulfide bond-dependent. Binding, albeit less, was also observed when the cysteine in domain II was substituted with an alanine, indicating that disulfide bonds may be involved in the stabilization of the apoE3/tau complex.

Hyperphosphorylated PHF-tau does not bind to tubulin or apoE3. The effects of phosphorylation on the binding of apoE3 to domain I, which contains the only phosphorylation site in the microtubule-repeat binding region, have also been examined (Huang et al. 1995). A synthetic 18 amino acid domain I peptide was phosphorylated at serine residue 162 and then incubated with human apoE3. Unlike the nonphosphorylated peptide, the phosphorylated peptide did not form a complex.

Real-Time Bimolecular Interaction Analyses

All of the interactions described above have examined irreversible interactions using western blotting techniques following denaturing polyacrylamide gel electrophoresis. Our laboratory is now investigating the initial reversible nature of these interactions using real-time bimolecular interaction analysis (BIA; Pharmacia Biosensor AB, Sweden), which will permit real time characterization of the binding between immobilized tau and free apoE.

The basis for measurements with real-time BIA is surface plasmon resonance (SPR), an optical phenomenon arising in thin metal films under conditions of total internal reflection (Fig. 1; BIAtechnology Handbook 1994). The phenomenon produces a sharp decrease in the intensity of reflected lights at a specific angle, the resonance angle. One molecule is immobilized on the sensor surface, a gold film coated glass slide, while a solution containing the other molecule flows continuously over the sensor surface. As molecules from the solution bind to the immobilized molecule, the resonance angle changes as a result of the increased molecular mass. A response is monitored by a fixed array of light-sensitive diodes, and a computer calculates the resonance angle.

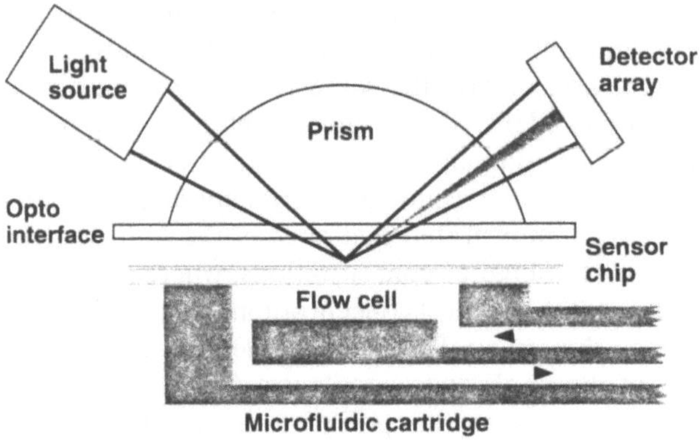

Fig. 1. The optical system and flow cell in real-time bimolecular interaction analysis (BIA) instruments (reprinted with permission from Pharmacia Biosensor AB, Uppsula, Sweden)

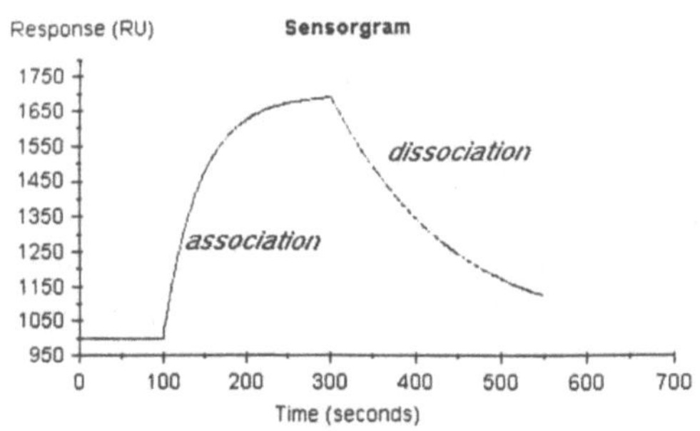

Fig. 2. Example of a simple sensorgram The *ordinate* shows the surface plasmon resonance (SPR) signal in resonance units, and the *abcissa* is real time in seconds (reprinted with permission from Pharmacia Biosensor AB, Uppsula, Sweden)

Results from real-time BIA are presented as a sensorgram, a plot of changes in the resonance signal as a function of time (Fig. 2). Progress of the binding is followed by observing changes in the resonance signal, and kinetic data are calculated from the rate of change of the signal. If the ordinate is the SPR signal and the abscissa is real time in seconds, a steadily increasing curve is generated as the molecule in solution binds to the immobilized molecule until equilibrium is reached. At equilibrium, the curve reaches a plateau. At this point, a ligand-free solution can be washed over the sensor chip and a resulting steadily decreasing curve reflects dissociation of the molecules. The two kinds of information that can be obtained from the sensorgram are the rate of interaction (association and dissociation) and the binding equilibrium.

The primary advantages of real-time BIA as an investigative technique are: (a) real-time analysis for obtaining kinetic data; (b) label-free detection; and (c) rapid interaction analysis on a time scale down to a few minutes per interaction. BIA is suitable for obtaining information on kinetics, affinity, specificity, concentration, cooperativity, and relative binding patterns of different protein species.

Pilot real-time BIA experiments on the binding of free apoE isoforms to immobilized tau indicate that the apoE3 and apoE4 isoforms bind to tau in a time-dependent and saturable manner. When apoE is eliminated from the solution in continuous flow over the immobilized tau, the complexes dissociate, indicating that the initial binding of the three isoforms to tau is reversible. One testable hypothesis is that each of the apoE isoforms can form a reversible complex with tau and that a small component of the apoE3/tau and apoE2/tau complexes proceeds to form an irreversible bond. These irreversible complexes are detectable by western blotting following denaturing poyacrylamide gel electrophoresis.

A Hypothesis for the Pathogenesis of Apolipoprotein E4 Gene-Related Alzheimer's Disease

The binding of apoE3 to tau may regulate the function and metabolism of tau (Strittmatter et al. 1994a). This regulation would not be present in individuals who only make apoE4 protein, or would be reduced in heterozygous individuals. We have proposed that individuals inheriting one or two APOE3 genes or an APOE2 gene, which decreases the risk and increases the age of onset of AD, can bind and sequester tau from forming PHFs or from phosphorylation, thus effectively protecting tau and slowing the formation of NFTs. The sequestered tau would be available in a conformation suitable for stabilizing microtubules. ApoE4 does not interact with tau in a similar manner, and therefore may not perform a protective or enhancing function. A compound designed to mimic the neuronal metabolic role of apoE3 or apoE2 may be a useful treatment to prevent or delay the onset of AD.

References

BIAtechnology Handbook (June 1994 edition) Pharmacia Biosensor AB, Uppsala, Sweden

Bramblett GT, Goedert M, Jakes R (1993) Abnormal tau phosphorylation at Ser^{396} in Alzheimer's disease recapitulates development and contributes to reduced microtubule binding. Neuron 10. 1089–1099

Corder EH, Sauders AM, Strittmatter WJ, Schmechel DE, Gaskell PC, Small GW, Roses AD, Haines JL, Pericak-Vance MA (1993) Gene dose of apolipoprotein E type 4 allele and the risk of Alzheimer's disease in late onset families. Science 261. 921–923

Diedrich JF, Minnigan H, Carp RI, Whitaker JN, Race R, Frey W II, Haase AT (1991) Neuropatholgical changes in scrapie and Alzheimer's disease are associated with increased expression of apolipoprotein E and Cathepsin D in astrocytes J Virol 65: 4759–4768

Goedert M (1993) Tau protein and the neurofibrillary pathology of Alzheimer's disease. Trends Neurosci 76 460–465

Goedert M, Spillantini MG, Cairns NJ, Crowther RA (1992) Tau proteins of Alzheimer paired helical filaments· abnormal phosphorylation of all six brain isoforms Neuron 8 159–168

Han S-H, Hulette C, Saunders AM, Einstein G, Pericak-Vance MA, Strittmatter WJ, Roses AD, Schmechel DE (1994a) Apolipoprotein E is present in hippocampal neurons without neurofibrillary tangles in Alzheimer's disease and in age-matched controls Exp Neurol 128 13–26

Han S-H, Einstein G, Weisgraber K, Strittmatter WJ, Saunders AM, Pericak-Vance MA, Roses AD, Schmechel DE (1994b) Apolipoprotein E is localized to the cytoplasm of human cortical neurons a light and electron microscopic study J Neuropathol Exper Neurol 53 535–544

Huang DY, Goedert M, Jakes R, Weisgraber KH, Garner CC, Saunders AM, Pericak-Vance MA, Schmechel DE, Roses AD, Stritmatter WJ (1994) Isoform specific interactions of apolipoprotein E with the microtubule-associated protein MAP2c implications for Alzheimer's disease Neurosci Lett 182 55–58

Huang DY, Weisgraber KH, Goedert M, Sauders AM, Strittmatter WJ (1995) ApoE3 binding to tau tandem repeat I is abolished by tau serine$_{262}$ phosphorylation Neurosci Lett 192. 1–4

Kounnas MZ, Morris RE, Thompson MR, Fitzgerald DJ, Strickland DK, Saelinger CB (1992) The α2-macroglobulin receptor/low density lipoprotein recepter related protein binds and internalizes Pseudomonas exotoxin A J Biol Chem 267 12420–12423

Lee VM-Y, Balin BJ, Otvos L, Trojanowski JQ (1991) A68 a major subunit of paired helical filaments and derivatized forms of normal tau Science 251 675–678

Namba Y, Tomonaga M, Kaqasaki H, Otomo E, Ikeda K (1991) Apolipoprotein E immunoreactivity in cerebral amyloid deposits and neurofibrillary tangles in Alzheimer's disease and kuru plaque amyloid in Creutzfeld-Jakob disease Brain Res 541 163–166

Saunders AM, Strittmatter WJ, Schmechel D, St George-Hyslop PH, Pericak-Vance MA, Joo SH, Rosi BL, Gusella JF, Crapper-Maclachan DR, Alberts MJ, Hulette C, Crain B, Boldgaber D, Roses AD (1993) Association of apolipoprotein E allele ε4 with late-onset familial and sporadic Alzheimer's disease Neurology 43 1467–1472

Schmechel DE, Sauders AM, Strittmatter WJ, Crain BJ, Hulette CM, Joo SH, Pericak-Vance MA, Goldgaber D, Roses AD (1993) Increased amyloid β-peptide deposition in cerebral cortex as a consequence of apolipoprotein E genotype in late-onset Alzheimer disease Proc Natl Acad Sci USA 90 9649–9653

Strittmatter WJ, Saunders AM, Schmechel DE, Pericak-Vance MA, Enghild J, Salvesen GS, Roses AD (1993) Apolipoprotein E high avidity binding to β-amyloid and increased frequency of type 4 allele in late-onset familial Alzheimer disease Proc Natl Acad Sci USA 90 1977–1981

Strittmatter WJ, Weisgraber KH, Goedert M, Saunders AM, Huang D, Corder EH, Dong L, Jakes R, Alberts MJ, Gilbert JR, Han S, Hulette C, Einstein G, Schmechel DE, Pericak-Vance MA, Roses AD (1994a) Hypothesis microtubule instability and paired helical filament formation in the Alzheimer disease brain are related to apolipoprotein E genotype Exp Neurol 125 163–171

Strittmatter WJ, Saunders AM, Goedert M, Weisgraber KH, Dong L, Jakes R, Huang D, Pericak-Vance MA, Schmechel DE, Roses AD (1994b) Isoform-specific interactions of apolipoprotein E with microtubule-associated protein tau implications for Alzheimer disease Proc Natl Acad Sci USA 91· 11183–11186

Weisgraber KH (1994) Apolipoprotein E structure-function relationships Adv Protein Chem 45 249–302

Wisniewski T, Frangione B (1992) Apolipoprotein E. a pathological chaperone protein in patients with cerebral and systemic amyloid Neurosci Lett 135· 235–238

Yoshida H, Ihara Y (1993) Tau in paired helical filaments is functionally distinct from fetal tau assembly incompetence of paired helical filament tau J Neurochem 61 1183–1186

Chaperoning Amyloid in Alzheimer's Disease: The Art of Avoiding Sticky Situations?

B. Frangione, E.M. Castaño, F. Prelli, C. Soto, J. Ghiso,* and *T. Wisniewski*

Abstract

Alzheimer's amyloid beta (Aβ) has long been known to be a highly insoluble and "sticky" protein. More recently it has become evident that several apolipoproteins (apo) can interact with Aβ. Apo J is a major carrier protein of normal soluble amyloid β (sAβ), which may well be the precursor of Aβ. On the other hand, apoE4 is linked to late-onset Alzheimer's disease and can modulate Aβ fibril formation in vitro. Furthermore, the carboxyl terminus of apoE is a constituent of Alzheimer's plaque amyloid. It is our working hypothesis that apoE, and possibly other chaperone proteins, may play a role in the conformational transition from soluble Aβ to amyloid fibrils.

Introduction

Alzheimer's disease (AD) is the most common illness characterized by the accumulation of amyloid and is the major cause of dementia (for review see Wisniewski et al. 1994a). Our work and that of other groups has recently shown the importance of apolipoproteins in the pathogenesis of AD (Wisniewski and Frangione 1992; Strittmatter et al. 1993a; Saunders et al. 1993; van Duijn et al. 1994; Wisniewski et al. 1994b, 1995a; Ma et al. 1994). The defining neuropathological lesions of AD are the deposition of amyloid β (Aβ) in the form of amyloid fibrils in congophilic angiopathy and senile plaques, as well as the aggregation of abnormally phosphorylated tau into neurofibrillary tangles. The Aβ contained in senile plaques is characterized by an altered conformational state that is predominantly β pleated, and by its resistance to proteolysis. However, it is increasingly being recognized that Aβ is not the only important component of senile plaques; indeed, our recent work has suggested that Aβ may not even be the only fibrillogenic protein in senile plaques (Wisniewski et al. 1995a,b).

---

* Department of Pathology, New York University Medical Center, 550 First Avenue, Room TH 427, New York, NY 10016, USA

A.D. Roses et al. (Eds.)
Apolipoprotein E and Alzheimer's Disease

$A\beta$ is normally present in body fluids as a soluble component (s$A\beta$) and is secreted by cells in culture (Shoji et al. 1992; Haass et al. 1992; Seubert et al. 1992; Busciglio et al. 1993). Amino acid sequence analysis indicated that the major form of s$A\beta$ is homologous to the amyloid protein extracted from cerebrovascular lesions, $A\beta$ 1–40 (Seubert et al. 1992). Further studies employing affinity chromatography, sequence analysis and laser desorption mass spectrometry identified minor s$A\beta$ sequences with heterogeneity and length similar to the $A\beta$ components deposited in senile plaques (Busciglio et al. 1993; Vigo-Pelfrey et al. 1993). The fact that this peptide has the same amino acid sequence as the $A\beta$ found in amyloid deposits and is able to cross the blood-brain barrier (Zlokovic et al. 1994, 1995) makes it likely that s$A\beta$ can contribute, at least partially, to the amyloid of AD. Whether or not s$A\beta$ is the immediate precursor of $A\beta$, it is clear that peptides with the same amino acid sequence can adopt both fibrillar and non-fibrillar conformations. Significantly, in vitro data on synthetic peptides homologous to s$A\beta$ have shown that these peptides form amyloid-like fibrils spontaneously (Castaño et al. 1986; Kirschner et al. 1987), and that normal biological fluids such as cerebrospinal fluid (CSF) inhibit this fibrillogenesis (Wisniewski et al. 1993). Our data suggest that apolipoprotein (apo) J is a major binding protein of the normal, non-fibrillar s$A\beta$ and that it inhibits the spontaneous polymerization of $A\beta$ peptides in vitro (Ghiso et al. 1993; Koudinov et al. 1994; Matsubara et al. 1995; Golabek et al. 1995). On the other hand, it is our working hypothesis that apoE and possibly other chaperone proteins may play an important role in the conformational transition of s$A\beta$ into amyloid fibrils.

Apolipoprotein J as the Carrier Protein for Soluble Amyloid β

We have recently identified what appears to be the major plasma and CSF carrier protein for s$A\beta$, ApoJ (Ghiso et al. 1993). Our data indicate that the complex of s$A\beta$-apoJ exists in normal CSF and that s$A\beta$ co-localized with apoJ containing high density plasma lipoproteins in plasma (Ghiso et al. 1993; Koudinov et al. 1994; Matsubara et al. 1995; Golabek et al. 1995). In vitro binding experiments indicate that the interaction between apoJ and s$A\beta$ has a very high affinity (K_d, 2 nM; Matsubara et al. 1995). The complex cannot be dissociated by the presence of physiologic concentrations of other plasma proteins with known $A\beta$ binding activity (e.g., apoE, α_1-antichymotrypsin, vitronectin or transthyretin; Matsubara et al. 1995).

In addition, apoJ may also be related to the delivery of s$A\beta$ through the blood–brain barrier (Zlokovic et al. 1994, 1996). Our recent in vivo studies performed in guinea pigs have demonstrated the existence of cerebrovascular permeability for $A\beta$ 1–40 and human apoJ as well as for the ApoJ–$A\beta$ 1–40 complex (Zlokovic et al. 1994, 1996). The association with apoJ enhances the transport of $A\beta$ 1–40 at the blood-brain barrier, through a receptor mediated mechanism that most likely involves gp 330/megalin, the receptor for ApoJ (Zlokovic et al. 1996).

ApoE as a Modulator of Fibrillogenesis

To test whether apoE functions as a modulator of fibrillogenesis of Aβ, we used an in vitro model. An assay was used where the amount of amyloid fibril formation in a solution was determined using fluorescence in the presence of Thioflavine T (Th-T; Wisniewski et al. 1994b). The studies were done at a concentration of 250μM Aβ 1–40 and 2.5μM apoE3 or apoE4. These concentrations are much higher than those found in biological fluids and were chosen to reflect the possible high levels found in Alzheimer's brain tissue. Aβ 1–40 was incubated alone or with recombinant apoE3 or E4 at room temperature for the indicated intervals. Fluorometry was performed as described elsewhere (Wisniewski et al. 1994b). During the 10 days of incubation there was a gradual increase of the fluorescence of both Aβ alone and Aβ co-incubated with apoE3 and E4. However, the rate of increase was greatest with Aβ 1–40 co-incubated with apoE4 (Fig. 1). The degree of fibril formation was also determined directly by electron microscopy of negatively stained amyloid fibrils. A far greater number of fibrils was observed in the presence of apoE4 (Fig. 2). Further fibrillogenesis studies were done using the Dutch variant Aβ 1–40-Q (Levy et al. 1990; Castaño et al. 1995). Similar variant peptides have been shown to be highly amyloidogenic in vitro, forming more stable fibrils and at a greater rate than Aβ 1–40 (Wisniewski et al. 1991). After incubation

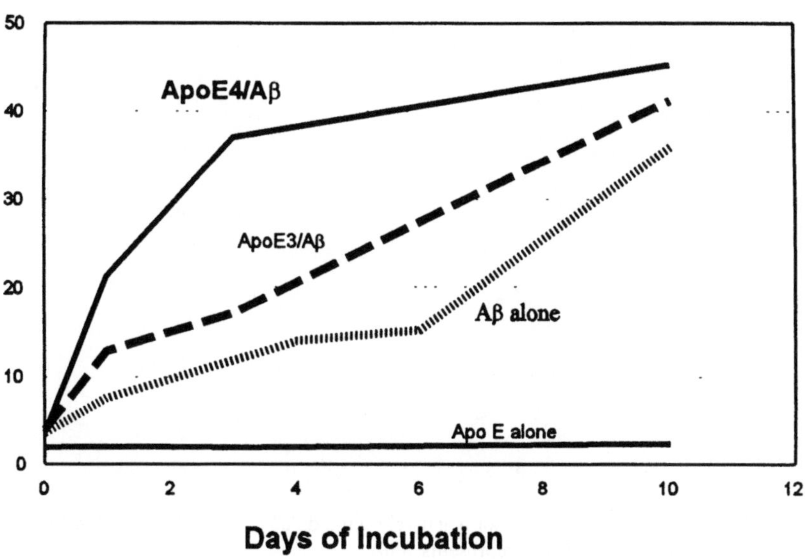

Fig. 1. Thıoflavıne T assay of Aβ 1–40 amyloıd fibrıl formatıon The mean fluorescence (ın arbıtrary unıts) ıs plotted agaınst the ıncubatıon perıod ın days of Aβ 1–40 wıth apoE4 (*top lıne*), Aβ 1–40 wıth apoE3 (*long dashes*), Aβ alone (*short dashes*) and apoE along (*bottom lıne*) In thıs assay, the amount of fluorescence ıs proportıonal to amyloıd fibrıl formatıon The greatment fluorescence was observed wıth Aβ 1–40 ın combınatıon wıth apoE4

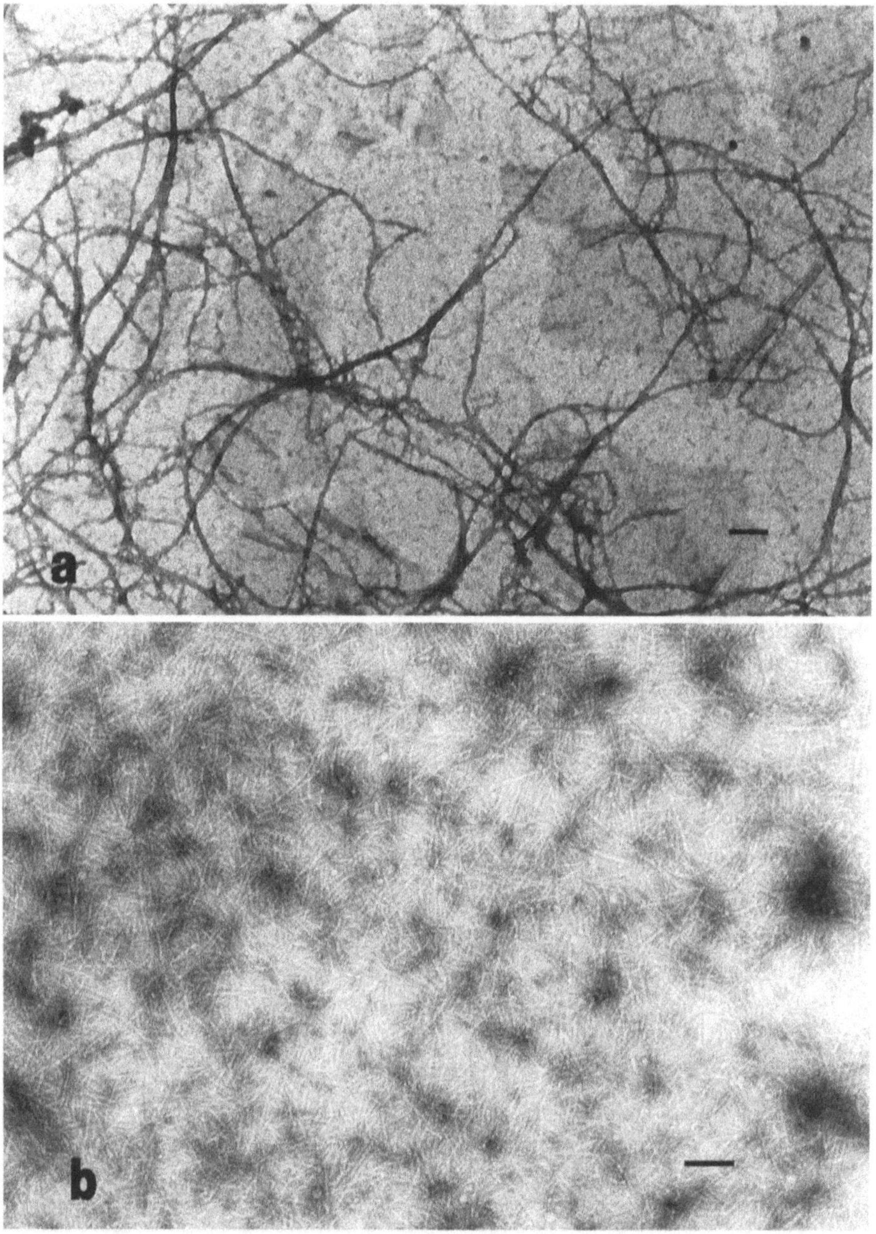

Fig. 2a,b. Electron micrography of negatively stained Aβ 1–40 fibrils (**a**) and Aβ 1–40 incubated with apoE4 (**b**) The *scale bar* represents 700Å for 24 h, Aβ 1–40Q formed large sheet aggregates that showed intense green birefringence after Congo Red staining. ApoE had minimal affect on the fibrillogenesis of this peptide. We further studied the effect of apoE on Th-T fluorescence monitored amyloid formation by using an Aβ analog that contains a replacement of Val for Ala at position 18 (Aβ 1–40A; Castaño et al. 1995). This peptide has a higher α-helix content, resulting in reduced fibrillogenic properties and a requirement for higher concentrations (600–800 μM) in order to display Congo Red birefringence after 24 h of incubation. When Aβ 1–40A was incubated for 24 h at 300 *μM*, no fluorescence above background levels was detected. Co-incubation with apoE (3.0 *μM*) resulted in a marked increase of fluorescence after 6 and 24 h. These results indicate that apoE produces a bigger effect on those peptides that have less β-sheet structure (Aβ 1–40A), consistent with the proposed idea that apoE promotes fibril growth and accelerates amyloid formation (Castaño et al. 1995). Whether these effects of apoE on Aβ fibrillogenesis, documented in vitro, actually occur in vivo is not known. A number of alternative roles for apoE in Aβ fibrillogenesis are also possible. For example apoE may bind preferentially to Aβ polymers and act to stabilize them. These and other possibilities are being addressed in experiments that represent in vivo conditions more closely.

The Carboxyl End of ApoE Is Complexed to Aβ

We have provided in vitro evidence for a possible Aβ/apoE interaction. In addition, we sought indications for such an interaction in vivo. Therefore we purified Aβ from senile plaques to determine if any apoE co-purified (Wisniewski et al. 1995a). Aβ purification was done using a modification of published protocols (Miller et al. 1993). The purification procedure was followed by Congo Red staining and Western blotting. Fractions containing Aβ and apoE fractions were subjected to amino acid sequencing and mass spectroscopy. Reverse phase high-performance liquid chromatography (RP-HPLC) of the purified amyloid yielded the following Aβ peptides: Aβ 1–42, Aβ 1–40, Aβ 4–42, Aβ 2–40, Aβ 8–42 and Aβ 11–42. In addition, a small peak eluted at the tail end of the Aβ peak on RP-HPLC, which corresponded to a 10 kDa C-terminal fragment of apoE by western blotting. The amino acid sequence showed a major start at residue 216 (Fig. 3) and a minor one at 193 of apoE. The yield of the carboxyl fragment of apoE, in comparison to Aβ, was approximately 1%–2%. This co-purification of apoE and Aβ suggests a close and high affinity interaction between these two proteins under the pathological conditions found in the AD brain. This finding has now been confirmed by another group (Naslund et al. 1995).

To evaluate the possible role of this fragment of apoE, in vitro studies were performed using thrombin-cleaved recombinant apoE. Ten mg of recombinant apoE3 was thrombin cleaved and separated as previously described.

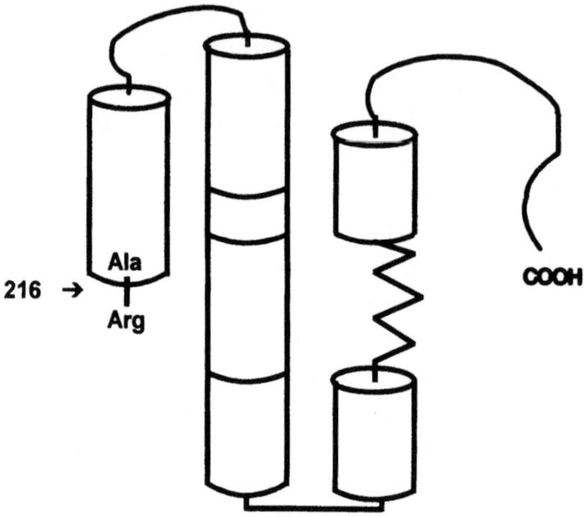

Fig. 3. The thrombin-generated carboxyl-terminal fragment of ApoE, which co-purified with amyloid β The major sequence began at residue 216

The C-terminal fragments, at 2 mg/ml in phosphate buffered saline (PBS), were incubated at room temperature for 10 days. The pelleted material was then subjected to negative staining and electron microscopy (EM). The latter revealed amyloid-like fibrils, which are seen in Fig. 4a. These fibrils were 8–15 nm in diameter and up to 500 nm in length and were either straight or twisted around each other. EM of the intact apoE revealed amorphous aggregates. Furthermore, the carboxyl-terminal apoE fragment and intact apoE, after a 10-day incubation at room temperature, were Congo Red stained (Fig. 4b). Intact apoE was Congo Red negative, whereas the C-terminus of ApoE showed the characteristic apple-green birefringence of amyloid (Wisniewski et al. 1995a). These studies have demonstrated that the carboxyl terminus of apoE can itself form amyloid-like fibrils in vitro that are Congo Red positive. Furthermore this fragment of apoE is found in vivo as a component of neuritic plaques.

Consistent with the above findings, it was previously documented that the binding site of apoE to Aβ is on its carboxyl terminus (Strittmatter et al. 1993b). Furthermore, the intact protein or a degradative fragment of several apolipoproteins is already known to have the ability to form amyloid, including apoAI, apoAII, and serum amyloid A protein (SAA; Ghiso et al. 1994). Hence, it is not surprising that a fragment of another apolipoprotein, apoE, can itself form amyloid.

Our results suggest that senile plaques may contain two types of amyloid fibrils. Amyloid fibril formation in vitro can be initiated by the addition of a template. The C-terminus of apoE, as well as the intact apoE, have the potential to initiate Aβ amyloid growth (Wisniewski et al. 1995a,b). In late-onset AD it is unclear which comes first, Aβ or apoE. The carboxyl terminal 190 to 299 residues of apoE is the same in E2, E3 and E4, with the amino acid differences between these isotypes being found in the N-terminal thrombolytic fragment. One possible way by which the presence of the carboxyl fragment of

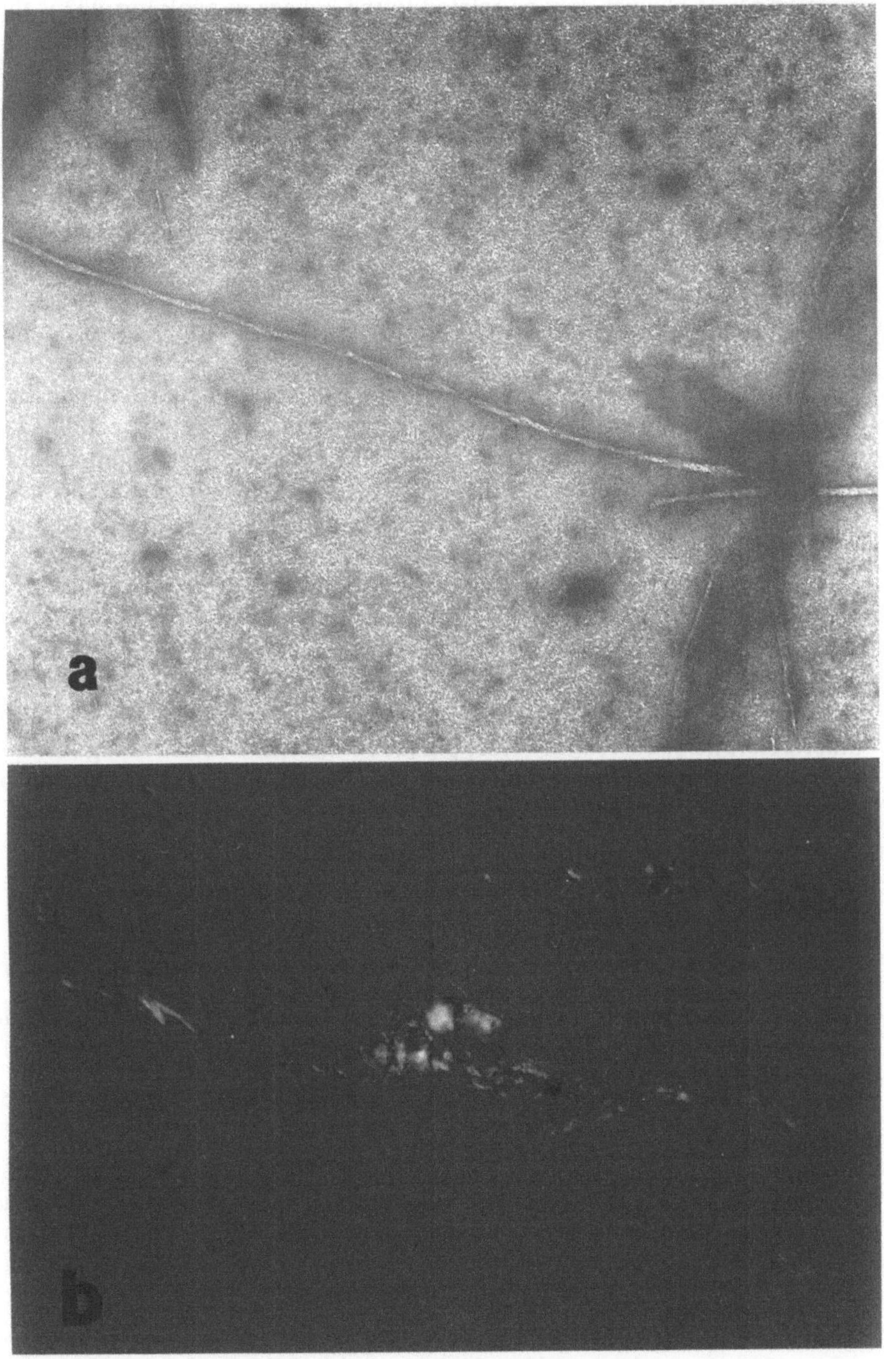

Fig. 4. a Electron micrograph of negatively stained carboxyl terminal fragment of ApoE, produced by the thrombin cleavage of recombinant ApoE Amyloid-like fibrils are present **b** Congo Red-stained carboxyl-fragment of ApoE, viewed under polarized light The characteristic apple-green birifringence of amyloid is present

apoE in senile plaques may be related to the linkage of apoE4 to AD is that the apoE4 isoform, being monomeric, has differing binding affinities and susceptibility to proteolytic attack. In an analogous manner certain SAA isotypes, such as SAA2, are known to be more likely to undergo degradation leading to AA amyloid formation (Ghiso et al. 1994).

We have added another member to the list of potentially fibrillogenic proteins in AD: apoE. Hence AD can be viewed as a disease defined neuropathologically by the presence of several interacting proteins that can adopt an amyloidogenic conformation. We suggest that some proteins, such as apoE or Aβ, when misfolded can propagate this abnormal conformation to other proteins, inducing them to fold into the same abnormal conformation. This "conformational mimicry" may initiate and/or augment fibrillogenesis in AD (Wisniewski et al. 1995b). This hypothesis of conformational mimicry has similarities to some of the proposed mechanisms for scrapie replication (Prusiner 1991), where it has been shown in vitro that a conversion of an α-helix to β-sheet conformation occurs using synthetic peptides homologous to fragments of PrP. Analogous to the prion-related diseases, we view AD as a disease characterized by abnormal protein conformation. However, unlike the prion diseases, we suggest that the conversion from a protein with a normal conformation (nac, non-amyloid conformation) to amyloid (ac, amyloid conformation; Soto and Frangione 1995) can be initiated by proteins that do not necessarily have the same amino acid sequence.

For many years it has been known that apolipoproteins are critical in atherosclerotic disease, with vessel plaque formation. It is becoming increasingly evident that apolipoproteins are also bound up in the production of another plaque, the senile plaque.

References

Busciglio J, Gabuzda D, Matsudaira P, Yanker B (1993) Generation of β-amyloid in the secretory pathway in neuronal and non neuronal cells Proc Natl Acad Sci USA 90 2092–2096

Castaño E, Ghiso J, Prelli F, Gorevic P, Migheli A, Frangione B (1986) In vitro formation of amyloid fibrils from two synthetic peptides of different lengths homologous to Alzheimer's disease β-protein Biochem Biophys Res Commun 141. 782–789

Castaño EM, Prelli F, Wisniewski T, Golabek A, Kumar RA, Soto C, Frangione B (1995) Fibrillogenesis in Alzheimer's disease of amyloid β peptides and apolipoprotein E Biochem J 306 599–604

Ghiso J, Matsubara E, Koudinov A, Choi-Miura N, Tomita M, Wisniewski T, Frangione B (1993) The cerebrospinal fluid soluble form of Alzheimer's amyloid beta is complexed to SP-40, 40 (apolipoprotein J), an inhibitor of the complement membrane attack complex Biochem J 293 27–30

Ghiso J, Wisniewski T, Frangione B (1994) Unifying features of systemic and cerebral amyloidosis Mol Neurobiol 8 49–64

Golabek A, Marques MA, Lalowski M, Wisniewski T (1995) Alzheimer's disease amyloid binding proteins in vitro and in normal human cerebrospinal fluid Neurosci Lett 191 79–82

Haass C, Schlossmacher M, Hung A, Vigo-Pelfrey C, Mellon A, Ostaszewski B, Lieberburg I, Koo E, Schenk D, Teplow D, Selkoe D (1992) Amyloid β-peptide is produced by cultured cells during normal metabolism Nature 359 322–325

Kirschner D, Inouye H, Duffy L, Sinclair A, Linda M, Selkoe D (1987) Synthetic peptides homologous to β-protein from Alzheimer's disease form amyloid-like fibrils in vitro. Proc Natl Acad Sci USA 84· 6953–6957

Koudinov A, Matsubara E, Frangione B, Ghiso J (1994) The soluble form of Alzheimer's amyloid beta protein is complexed to high density lipoprotein 3 and very high density lipoprotein in normal human plasma Biochem Biophys Res Commun 205 1164–1170

Levy E, Carman MD, Fernandez-Madrid IJ, Power MD, Lieberburg I, VanDuinen SG, Bots GTAM, Luyendijk W, Frangione B (1990) Mutation of the Alzheimer's disease amyloid gene in hereditary cerebral hemorrhage, Dutch type Science 248 1124–1126

Ma J, Yee A, Brewer HB, Das S, Potter H (1994) The amyloid associated proteins α_1 antichymotrypsin and apolipoprotein E promote the assembly of the Alzheimer's β-protein into filaments Nature 372 92–94

Matsubara E, Frangione B, Ghiso J (1995) Characterization of apolipoprotein J-Alzheimer's Aβ interaction J Biol Chem 270 7563–7567

Miller DL, Papayannopoulos LA, Styles J, Bobin SA, Lin YY, Bieman K, Iqbal K (1993) Peptide composition of the cerebrovascular and senile plaque core amyloid deposits of Alzheimer's disease Arch Bioch Biophys 301 41–52

Naslund J, Thyberg J, Tjernberg LO, Wernstedt C, Karlstrom AR, Bogdanovic N, Gandy SA, Lanndelt L, Terenius L, Nordstedt C (1995) Characterization of stable complexes involving apolipoprotein E and the amyloid β peptide in Alzheimer's disease brain Neuron 15 219–228

Prusiner S (1991) Molecular biology of prion diseases. Science 252 1515–1522

Saunders AM, Strittmatter WJ, Schmechel D, St George-Hyslop PH, Pericak-Vance MA, Joo SH, Rosi BL, Gusella JF, Grapper-Maclachan DR, Alberts MJ, Hulette C, Crain B, Goldgaber D, Roses AD (1993) Association of apolipoprotein E allele $\varepsilon 4$ with late-onset familial and sporadic Alzheimer's disease Neurology 43 1467–1472

Seubert P, Vigo-Pelfrey C, Esch F, Lee M, Dovey H, Davis D, Sinha S, Schlossmacher M, Whaley J, Swindlehurst C, McCormack R, Wolfert R, Selkoe D, Lieberburg I, Schenk D (1992) Isolation and quantification of soluble Alzheimer's β-peptide from biological fluids Nature 359 325–327

Shoji M, Golde T, Ghiso J, Cheung T, Estus S, Shaffer L, Cai X-D, McKay D, Tintner R, Frangione B, Younkin S (1992) Production of the Alzheimer's amyloid β protein by normal proteolytic processing Science 258 126–129

Soto C, Frangione B (1995) Two conformational states of amyloid β-peptide implications for the pathogenesis of Alzheimer's disease Neurosci Lett 186 115–118

Strittmatter WJ, Saunders AM, Schmechel D, Pericak-Vance MA, Enghild J, Salvesen GS, Roses AD (1993a) Apolipoprotein E high-avidity binding to β-amyloid and increased frequency of type 4 allele in late-onset familial Alzheimer's disease Proc Natl Acad Sci USA 90 1977–1981

Strittmatter WJ, Weisgraber KH, Huang DY, Dong LM, Salvesen GS, Pericak-Vance M, Schmechel D, Saunders AM, Goldgaber D, Roses AD (1993b) Binding of human apolipoprotein E to synthetic amyloid β isoform specific effects and implications for late-onset Alzheimer's disease Proc Natl Acad Sci USA 90 8098–8102

van Dujin CM, de Knijff P, Cruts M, Wehnert A, Havekes LM, Hofman A, Van Broeckhoven C (1994) Apolipoprotein E4 allele in a population-based study of early-onset Alzheimer's disease Nature Genet 7 74–78

Vigo-Pelfrey C, Lee D, Keim P, Lieberburg I, Schenk D (1993) Characterization of β-amyloid peptide from human cerebrospinal fluid. J Neurochem 61 1965–1968

Wisniewski T, Frangione B (1992) Apolipoprotein E a pathologic chaperone protein in patients with cerebral and systemic amyloid. Neurosci Lett 135 235–238

Wisniewski T, Ghiso J, Frangione B (1991) Peptides homologous to the amyloid protein of Alzheimer's disease containing a glutamine for glutamic acid substitution have accelerated amyloid fibril formation Biochem Biophys Res Commun 179 1247–1254

Wisniewski T, Castaño E, Ghiso J, Frangione B (1993) Cerebrospinal fluid inhibits Alzheimer β-amyloid fibril formation in vitro Ann Neurol 34 631–633

Wisniewski T, Ghiso J, Frangione B (1994a) Alzheimer's disease and soluble Aβ Neurobiol Aging 15 143–152

Wisniewski T, Castaño EM, Golabek A, Vogel T, Frangione B (1994b) Acceleration of Alzheimer's fibril formation by apolipoprotein E in vitro Am J Pathol 145 1030–1035

Wisniewski T, Lalowski M, Golabek A, Vogel T, Frangione B (1995a) Is Alzheimer's disease an apolipoprotein E amyloidosis? The Lancet 345 956–958

Wisniewski T, Golabek AA, Kida E, Wisniewski KE, Frangione B (1995b) Conformational mimicry in Alzheimer's disease Am J Pathol 147: 238–244

Zlokovic BV, Martel CL, Mackic JB, Matsubara E, Wisniewski T, McComb JG, Frangione B, Ghiso J (1994) Brain uptake of circulating apolipoproteins J and E complexed to Alzheimer's amyloid β. Biochem Biophys Res Commun 205 1431–1437

Zlokovic BV, Martel CL, Matsubara E, McComb JG, Zneng G, McCluskey RT, Frangione B, Ghiso J (1996) Gp 330/megalin probable role in receptor-mediated transport of Apolipoprotein J alone and in complex with Alzheimer's amyloid β at the blood-brain and blood-cerebrospinal fluid barriers Proc Natl Acad Sci USA (in press)

Apolipoprotein E Oxidation and Functional Consequences

*B. Leininger-Muller**, *C. Jolivalt, T. Pillot, P. Lagrange, M.H. Livertoux, M.C. Grassiot, A. Minn*, and *G. Siest*

Introduction

Many theories have been advanced to account for the ageing process. Among them, the free radical theory of ageing remains the most generally accepted (Emerit and Chance 1992). The brain is believed to be particularly vulnerable to oxidative damage, as it contains high concentrations of easily peroxidizable lipids (i.e., those rich in polyunsaturated fatty acids), has regions particularly enriched in iron, and has comparatively low to moderate levels of endogenous antioxidant molecules (Halliwell 1992; Harris 1992). The activity of antioxidant enzymes like superoxide dismutase and catalase is increased in Alzheimer's disease (AD) brain (Balazs and Leon 1994; Palmer and Burns 1994). Protein oxidation products are reportedly increased in both aged rodent and human brains (Fucci et al. 1983; Smith et al. 1991), as well as in Alzheimer's patients (Volicer and Crino 1990). Measurements on postmortem brain samples have effectively shown that, compared with young controls, samples from aged groups have an increased carbonyl content and decreased glutamine synthetase and creatine kinase activities, two enzymes that are efficiently inhibited by radical oxygen species. These modifications are more marked in the frontal than in the occipital region. This suggests that, whereas protein oxidation products accumulate in the brain, oxidation-sensitive enzyme activities decrease with ageing in the same regional pattern (Smith et al. 1991). Overexpression of the cytochrome-oxidase gene has also been shown by Northern blot analysis in AD patients (Alberts et al. 1992).

Other enzymes are sensitive to free radicals. For instance, microsomal glutathione *S*-transferases are activated by hydrogen peroxide, associated with the formation of a protein dimer (Aniya and Anders 1992). Moreover, it has been shown that the increase in glutathione *S*-transferase activity in rat liver microsomes after phenobarbital treatment is the result of an oxidative modification of the enzyme by reactive oxygen species, whose production is concomitantly increased by the induction of cytochrome P450 (Aniya et al. 1993).

* Centre du Médicament, Université Henri Poincaré, CNRS URA 597, 30 Rue Lionnois, 54000 Nancy and Centre de Médecine Préventive, 2 Rue du Doyen Jacques Parisot, 54500 Vandoeuvre, France

A D Roses et al (Eds)
Apolipoprotein E and Alzheimer's Disease

Some drug-metabolizing enzymes also show a regulatory sequence in their gene: rat glutathione *S*-transferase Ya subunit gene contains a regulatory region, the antioxidant responsive element, that can be activated by phenolic antioxidants and hydrogen peroxide (Nguyen and Pickett 1992).

There are four main sources of oxygen-centered radical production in cells (Ames et al. 1993): (1) those of mitochondrial respiratory chain by-products, which are $O_2^{\bullet -}$, H_2O_2 and $OH^{\bullet}$; (2) phagocytic cells that destroy bacteria or virus-infected cells; (3) peroxisomes that are responsible for the degradation of fatty acids and other molecules and produce H_2O_2 as a by-product, which is then degraded by catalase; and (4) enzyme activities related to cytochrome P450 family that produce oxidant by-products, mainly superoxide and hydroxyl radicals, that may damage cell macromolecules.

For more than 10 years, we have studied cerebral drug metabolizing enzyme families and we have shown the heterogeneous regional and subcellular distribution of their specific activities in the rat brain. We have already shown that the reductive metabolism by monoelectronic transfer of some xenobiotics by brain microsomal reduced nicotinamide-adenine-dinucleotide phosphate (NADPH)-cytochrome P450 (*c*) reductase leads to the production of superoxide radicals (Ghersi-Egea et al. 1991). By using this reductive metabolism for the production of superoxide amounts comparable to those produced in vivo, we obtained a model that is a useful tool for the study of the action of these reactive species on other molecules present in the brain. Following some preliminary observations that suggested an inhibitory effect of superoxide radical on some cytochrome P450 activities, we used this model to study the oxidative modifications of two functional proteins, cytochrome P450 isoform CYP 1A1 and apolipoprotein E (apoE). Among the apolipoproteins present in the brain, apoE, the main apolipoprotein synthesized in the brain (Mahley 1988), provides a potential target for oxidative modification. Moreover, the implication of polymorphic expression of apoE in the etiology of AD and its presence in amyloid plaques are now well documentated (Wisniewski and Frangione 1992; Strittmatter et al. 1993a). In addition, some studies on the binding of apoE with Aβ peptide suggest that the oxidation of apoE should enhance its affinity to Aβ (Strittmatter et al. 1993b). In the same way, Aβ could also be oxidized. The neurotoxicity of Aβ has been claimed to be related to its capacity to aggregate and produce free radicals (Hensley 1994). Snyder et al. (1994) demonstrated that oxidation of Met-35 to the sulfoxide in Aβ-40 enhances the aggregation rate as compared to the non-oxidized peptide. Oxidative stress also increases the secretion and release of the amyloid peptide products (Gabuzda et al. 1994).

The first aim of our study was to establish whether brain NADPH-cytochrome P450 (*c*) reductase mediated superoxide radical formation might lead to the oxidation of apoE and what the consequences of this attack would be (Visvikis et al. 1994).

Second, a previous study by our group localized, in the human brains, most of the drug metabolizing enzyme activities previously described in the rat

(Ghersi-Egea et al. 1993). However, no study has determined whether there are modifications of the rates of drug metabolizing enzyme activities in AD brains, as compared to normal samples. As a result of observations in the literature that show a general overproduction of free radical generating systems in AD brains, either an eventual increase of drug metabolizing activities able to produce superoxide radicals as by-products or, inversely, a decrease of protective activities like epoxide hydrolases will have to be verified.

We present here the first demonstration that NADPH-cytochrome P450 (*c*) reductase mediated superoxide radical production is involved in apoE oxidation, leading to a dimerisation of this apolipoprotein. We also measured a cytochrome P450-dependent monooxygenase activity and NADPH-cytochrome P450 (*c*) reductase and microsomal epoxide hydrolase activities in several AD brain samples. The main results are described and discussed.

Methods

Studies on Human Tissue Preparation

Human brain tissue was obtained from the Kathleen Price Bryan Brain Bank (Duke University, NC, USA). The autopsy samples were from 12 subjects diagnosed clinically and neuropathologically as having AD and four subjects with no clinical or morphological evidence of AD (Table 1). Both groups had a similar postmortem delay (3.0 ± 1.3 h) and storage time at −80°C (2.0 ± 1.0

Table 1. Main characteristics of the autopsy samples obtained from the K Price Bryan Brain Bank

Patient number	Age (years)	Postmortem delay (h)	Diagnosis
437	83	3 5	AD
436	60	0 5	AD
572	67	3 0	AD
425	102	5 0	N
545	76	3 5	AD
543	72	3.0	N
488	70	2 75	AD
453	83	?	AD
406	77	1 0	N
550	60	5.5	AD
394	91	3.5	AD
511	72	4 0	AD
523	87	2 5	AD
560	69	3.5	AD
372	88	1 5	AD
542	81	3.25	N

N, not demented, AD, Alzheimer's disease

years). Two regions were investigated from each frozen brain: frontal cortex and cerebellum. The pial membranes, superficial blood vessels and most of the white matter were removed by dissection from the thawed tissues. The samples were homogenized in an ice-cold isolation medium (sucrose, 0.32 *M*; Tris, 10 m*M*; ethylenediaminetetra-acetic acid, EDTA, 1 m*M*, pH 7.4) by the use of a manual Dounce glass homogenizer, which prevents the contamination of the microsomal fraction by membranes of mitochondrial origin. Microsomal fractions were prepared as previously described (Ghersi-Egea et al. 1993).

Enzymatic Assay. In the human samples, NADPH-cytochrome P450 (*c*) reductase and 7-benzoxyresorufin *O*-dealkylase activities were measured as previously described (Ghersi-Egea et al. 1993). Protein was measured by the method of Lowry et al. (1951) using bovine serum albumin as a standard.

Measurement of Oxidation Products

Rat brain microsomes were prepared as previously described (Ghersi-Egea et al. 1991). The measurement of cytochrome P450-dependent activity in microsomes was carried out by a sensitive fluorimetric procedure using 7-benzoxyresorufin as substrate. For the measurement of the cytochrome P450-dependent activity after oxidation by superoxide radicals produced by the redox cycling of menadione, 1 m*M* menadione was added into the assays. After different incubation times, the reaction was started by the addition of a NADPH-regenerating system and was run at 37°C for 20 min.

Oxidation by Rat Brain Microsomes. Purified recombinant apoE produced by *Escherichia coli* was incubated with rat brain microsomes, with menadione in Tris buffer (50 m*M* Tris, 1 m*M* EDTA, 0.15 *M* NaCl, 20 m*M* urea, pH 7.4), in a final volume of 20 μl. Proteins were derivatized with an equal volume of 0.5 m*M* 2,4-dinitrophenyl hydrazine in 0.1 *M* sodium phosphate buffer (pH 6.3), following the addition of H_2O_2m*M*, and were incubated for 1 h at room temperature (Keller et al. 1993). Oxidized apoE was separated electrophoretically under reducing conditions with sodium dodecyl sulfate/polyacrylamide gel electrophoresis (SDS/PAGE) and transferred to PVDF membrane. Then, the membrane was incubated overnight at room temperature, with anti-dinitrophenyl antibodies (Sigma, diluted 1/1000), followed by a 1.5-h incubation with anti-rabbit immunoglobulin G (IgG) conjugated with alkaline phosphatase (Jackson Immuno Research, diluted 1/5000). The membrane was then developed with 5-bromo-4-chloro-3-indolyl phosphate/Nitroblue Tetrazolium (Sigma).

Metal-Catalyzed Oxidation. In another set of experiments, purified apoE was also oxidized by the vanadyl/hydrogen peroxide system generating hydroxyl radicals. Each sample contained 5 μg apoE, 1 m*M* H_2O_2 and the indicated amount of vanadyl sulfate in Tris buffer (pH 7.4). Proteins were derivatized with the equal volume of 0.5 m*M* 2,4-dinitrophenyl hydrazine in 0.1 *M* sodium phosphate buffer (pH 6.3), following the addition of H_2O_2, and were incubated for 1 h at room temperature (Keller et al. 1993). Oxidized apoE was separated electrophoretically under reducing conditions by SDS/PAGE and transferred to PVDF membrane. Then, the membrane was incubated overnight at room temperature, with anti-dinitrophenyl antibodies (Sigma, diluted 1/1000), followed by a 1-h incubation with anti-rabbit IgG conjugated with horseradish peroxidase diluted 1/3000. The membrane was then developed by Enhanced Chemiluminescence (ECL, Amersham).

Results

Studies on Human Brains

Because our previous studies showed that the cytochrome P450-dependent enzymatic activities towards xenobiotics exhibited a microsomal prevalence in human brain, we measured them only on this fraction (Table 2). Despite the large interindividual variations, these results confirm the two main observations made in our previous study (Ghersi-Egea et al. 1993): a low activity of NADPH-cytochrome P450 (*c*) reductase and a high specific activity of the membrane-bound epoxide hydrolase. For each region, there was no significant difference between the activities measured in AD brains as compared to controls. However, it must be kept in mind that the number of controls in this study is low.

Oxidation Studies

The first possible target of superoxides produced in the model that we studied was cytochrome P450 activity. The slow, time-dependent decrease of the enzymatic activity observed in both hepatic and brain samples may be the result of a minor production of superoxide directly mediated by the cytochrome P450 activity, which lead to an autooxidation of the enzymatic protein. The addition of menadione to the incubation medium resulted in an important decrease of the enzymatic activity (Fig. 1), which can be directly related to the

Table 2. Drug-metabolizing enzyme activities in human brain microsomes

	Reductase (nmol/min per mg protein)		BROD (pmol/min per mg protein)		mEH (pmol/min per mg protein)	
Regions	FC	Cerebellum	FC	Cerebellum	FC	Cerebellum
Controls	1.54 ± 1.2	0.68 ± 0.5	0.71 ± 0.1	0.70 ± 0.5	355.6 ± 26	365.0 ± 39
AD	1.54 ± 1.06	0.76 ± 0.5	0.72 ± 0.5	0.66 ± 0.2	401.6 ± 99	350.3 ± 66

Reductase, reduced nicotinamide-adenine-dinucleotide phosphate (NADPH)-cytochrome P450 (*c*) reductase activity, mEH, membrane-bound epoxide hydrolase activity, BROD, 7-benzoxyresorufin *O*-dealkylase activity, FC, frontal cortex

The data reported are the means ± SD of controls (n = 4) and Alzheimer's disease (AD) patients (n = 12) Individual data were the mean of two or three determinations

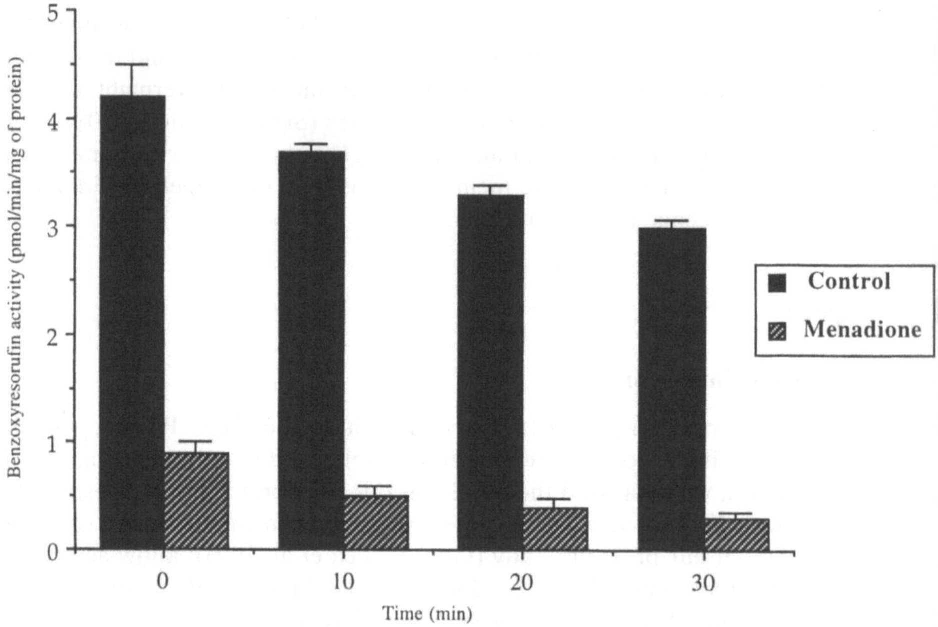

Fig. 1. Cytochrome P450 activities in rat brain microsomes Effect of oxidation on benzoxyresorufin activity by redox cycling of menadione (m*M*)

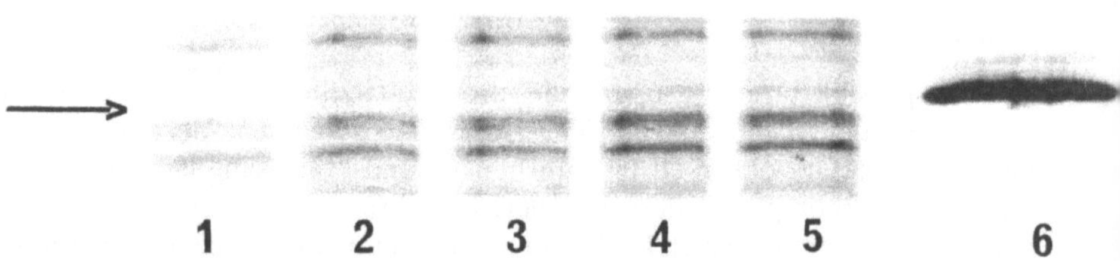

Fig. 2. Immunoblot showing the oxidation of recombinant apoE by rat brain microsomes, detected by the phosphatase alkaline substrate *Lane 1*, microsomes alone (40 μg), *lane 2*, microsomes plus apoE (4 μg), *lane 3*, microsomes plus apoE plus reduced nicotinamide-adenine-dinucleotide phosphate (NADPH)-regenerating system, *lane 4*, microsomes plus apoE plus NADPH-regenerating system plus 10 μM menadione, *lane 5*, microsomes plus apoE plus NADPH-regenerating system plus 100 μM menadione, *lane 6*, apoE (0 5 μg, detected by anti-apoE antibodies)

superoxide amount produced during the NADPH-cytochrome P450 (*c*) reductase redox cycling of the menadione.

ApoE represents the second potential target for brain-produced superoxides that we studied. Figure 2 shows that apoE can be effectively oxidized by superoxide radicals produced through NADPH-cytochrome P450 (*c*) reductase dependent pathways. The oxidation profile observed when recombinant apoE is oxidized by the chemical system reveals a band at 38 kDa when apoE is incubated without the chemical system, thus showing that the recombinant apoE that we used in our experimental conditions is already partially oxidized. However, the increase of superoxide radicals concentration in the medium lead to the appearance of a 66-kDa band, suggesting a dimerisation of apoE under these stronger oxidation conditions (Fig. 3).

Discussion

The measurements of the specific activities of the three drug-metabolizing enzymes do not show any significant difference between the AD brains and the controls. However, since we do not have enough control brain samples, studies are presently underway with supplementary samples. We are also looking at other drug metabolizing enzyme activities to see if there are modifications dependent on other parameters, such as apoE genotype and age of the patients, or related to the histological characteristics of the autopsy samples. The results obtained during the oxidation studies are of particular interest since they reveal two important facts. First, they demonstrate that apoE can be oxidized by superoxide radicals produced through NADPH-cytochrome P450 (*c*) reductase reductive pathway in the brain. Second, the formation of dimers of apoE as a consequence of its oxidation is of great interest. However, it remains to be verified if it is really due to the superoxide radicals produced or if it is the vanadyl that reacts with apoE to form aggregates.

In addition to oxidation, other protein modifications such as glycation should be taken into account (Kimura et al. 1995). ApoE glycation has been described as it is the case for other apolipoproteins in diabetic patients. Protein glycation starts with the non-enzymatic condensation of a sugar aldehyde or

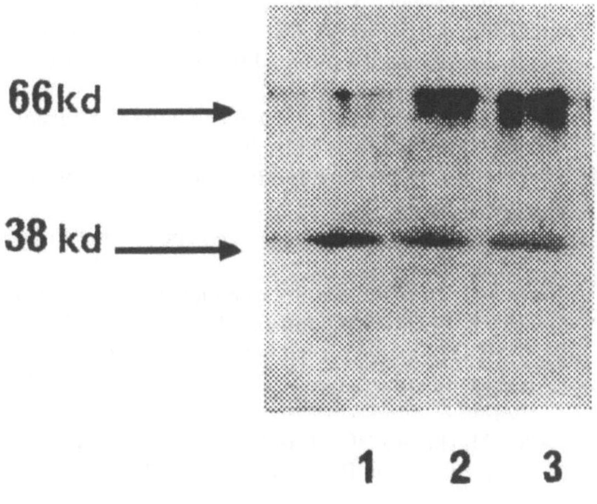

Fig. 3. Immunoblot showing metal-catalyzed oxidation of apoE using vanadyl/hydrogen peroxide system, detected by chemiluminescence *Lane 1*, non-oxidized apoE, *lane 2*, apoE plus 400 μM vanadyl, *lane 3*, apoE plus 800 μM vanadyl

ketone group with free amino groups, such as those of lysine or arginine to form a labile Schiff base, a reaction first described by Maillard (1912). The resulting adducts, after different rearrangements, could be found in senile plaques. These products form an heterogeneous group and oxidation interferes in the formation of these so-called AGEs (advanced glycosylation end products; Smith et al. 1995). Thus, oxidation should be studied simultaneously with glycation for a better understanding of the mechanisms leading to the "glyco-oxidation" products found in the brains of AD patients.

Acknowledgments. This work is supported in part by a grant from the IPSEN Foundation. The authors wish to thank Dr. A. Roses for providing autopsy human brain samples from the Kathleen Price Bryan Brain Bank.

References

Alberts MJ, Ioannou P, Deucher R, Gilbert J, Lee J, Middleton L, Roses A (1992) Isolation of a cytochrome oxidase gene overexpressed in Alzheimer's disease brain Mol Cell Neurosci 3 461–470

Ames BN, Shigenaga MK, Hagen TM (1993) Oxidants, antioxidants and the degenerative diseases of aging Proc Natl Acad Sci USA 90 7915–7922

Aniya Y, Anders MW (1992) Activations of rat liver microsomal glutathione S-transferase by hydrogen peroxide role for protein-dimer formation Arch Biochem Biophys 296 611–616

Aniya Y, Shimoji M, Naito A (1993) Increase in liver microsomal glutathione S-transferase activity by phenobarbital treatment of rats Biochem Pharmacol 46 1741–1747

Balazs L, Leon M (1994) Evidence of an oxidative challenge in the Alzheimer's brain Neurochem Res 19 1131–1137

Emerit I, Chance B (1992) Free radicals and aging Birkhauser, Basel

Fucci L, Oliver CN, Coon MJ, Stadtman ER (1983) Inactivation of key metabolic enzymes by mixed-function oxidation reactions possible implication in protein turnover and ageing Proc Natl Acad Sci USA 88 10540–10543

Gabuzda D, Busciglio J, Chen LB, Matsudaira P, Yankner BA (1994) Inhibition of energy metabolism alters the processing of amyloid precursor protein and induces a potentially amyloidogenic derivative J Biol Chem 269 13623–13628

Ghersi-Egea JF, Livertoux MH, Minn A, Perrin R, Siest G (1991) Enzyme mediated superoxide radical formation initiated by exogenous molecules in rat brain preparations. Toxicol Appl Pharmacol 110 107–117

Ghersi-Egea JF, Perrin R, Leininger-Muller B, Grassiot MC, Jeandel C, Floquet J, Cuny G, Siest G, Minn A (1993) Subcellular localization of cytochrome P450, and activities of several enzymes responsible for drug metabolism in the human brain Biochem Pharmacol 45 647–658

Halliwell B (1992) Reactive oxygen species and the central nervous system J Neurochem 59 1609–1623

Harris ED (1992) Regulation of antioxidant enzymes FASEB J 6 2675–2683

Hensley K, Carney JM, Mattson MP, Aksenova M, Harris M, Wu JF, Floyd RA, Butterfield DA (1994) A model for beta-amyloid aggregation and neurotoxicity based on free radical generation by the peptide relevance to Alzheimer disease Proc Natl Acad Sci USA 91 3270–3274

Keller RJ, Halmes NC, Hinson JA, Pumford NR (1993) Immunochemical detection of oxidized proteins Chem Res Toxicol 6 430–433

Kimura T, Takamatsu J, Araki N, Goto M, Kondo A, Miyakawa T, Horiuchi S (1995) Are advanced glycation end-products associated with amyloidosis in Alzheimer's disease? Neuroreport 6 866–868

Lowry OH, Rosebrough NJ, Farr AL, Randall RJ (1951) Protein measurement with the Folin phenol reagent J Biol Chem 193· 265–275

Mahley RW (1988) Apolipoprotein E cholesterol transport protein with expanding role in cell biology Science 240: 622–630

Maillard LC (1912) Action des acides aminés Sur des sucres Formation des melanoides par voie méthodique. CR Acad Sci 154. 66–68

Nguyen T, Pickett C (1992) Regulation of rat glutathione S-transferase Ya subunit gene expression J Biol Chem 267 13535–13539

Palmer AM, Burns MA (1994) Selective increase in lipid peroxidation in the inferior temporal cortex in Alzheimer's disease Brain Res 645 338–342

Smith CD, Corney JM, Starke-Reed PE, Oliver CN, Stadtman ER, Floyd RA, Markesbery WR (1991) Excess brain protein oxidation and enzyme dysfunction in normal aging and Alzheimer disease Proc Natl Acad Sci USA 88 10540–10543

Smith MA, Sayre LM, Monnier V, Perry G (1995) Radical AGEing in Alzheimer's disease Trends Neurosci 18 172–176

Snyder SW, Ladror US, Wade WS, Wang GT, Barrett LW, Matayoshi ED, Huffaker HJ, Krafft GA, Holzman TF (1994) Amyloid-beta aggregation selective inhibition of aggregation in mixtures of amyloid with different chain lengths. Biophys J 67 1216–1228

Strittmatter WJ, Saunders AM, Schmechel D, Pericack-Vance M, Enghild J, Salvesen GS, Roses AD (1993a) Apolipoprotein E high-avidity binding to β-amyloid and increased frequency of type 4 allele in late-onset familial Alzheimer disease Proc Natl Acad Sci USA 90· 1977–1981

Strittmatter WJ, Weigraber KH, Huang DY, Dong L, Salvesen GS, Pericack-Vance M, Schemechel D, Saunders AM, Goldgaber D, Roses AD (1993b) Binding of human apolipoprotein E to synthetic amyloid β peptide isoform-specific effects and implications for late-onset Alzheimer disease. Proc Natl Acad Sci USA 90: 8098–8102

Visvikis S, Leininger-Muller B, Lecomte E, Pillot T, Aguillon D, Artur Y, Bailly K, Bohnet K, Derex D, Galteau MM, Gueguen R, Herbeth B, Lagrange PH, Livertoux MH, Marangon K, Minn A, Sabolovic N, Steinmetz J, Visvikis A, Siest G, Blain H, Jeandel C, David F, Lucotte G apolipoprotéine E, oxydation et maladie d'Alzheimer In. Poncet M, Michel B, Nieoullon A (eds) Actualités sur la maladie d'Alzheimer et les syndromes apparentés. Solal, Marseille, pp 39–50

Volicer L, Crino PB (1990) Involvement of free radicals in dementia of the Alzheimer type a hypothesis Neurobiol Aging 11 567–571

Wisniewski T, Frangione B (1992) Apolipoprotein E a pathological chaperone protein in patients with cerebral and systemic amyloid Neurosci Lett 135· 235–238

Ethnic Variation in the Association Between APOE-ε4 and Alzheimer's Disease: A Follow-up Study

G. Maestre, P. Schofield, M. Chun, M.-X. Tang, B. Tycko,* and *R. Mayeux*

Summary

Apolipoprotein-E (APOE) genotypes with the ε4 allele have been associated with Alzheimer's disease (AD) and confirmed worldwide. We previously reported a diminished association between APOE-ε4 and AD among African-Americans compared to Caucasians and Hispanics in New York. In this report we used an expanded cohort of cases and controls from three ethnic groups in the same New York City community to further examine the association between APOE and AD.

The allelic frequency of APOE-ε4 was significantly higher in African-Americans compared to Caucasians and Hispanics, and this difference was most striking among older individuals. We found an equally strong association between APOE-ε4 homozygosity and AD among African-Americans, Hispanics and Caucasians. However, although there was a association between AD and APOE-ε4 heterozygosity in Caucasians and Hispanics, among African-Americans there was an inverse association. APOE-ε2 genotypes were associated with reduced risk in Caucasians, but no such relationship existed in African-Americans or Hispanics. Variability in the strength and type of association between AD and the APOE genotypes in the three ethnic groups could not be explained by age differences.

These results emphasize differences in the relationship between APOE and AD among different ethnic groups. It is likely that other genetic or environmental factors modify the APOE-ε4 effect in heterozygote African-Americans.

Introduction

The association between APOE-ε4 and Alzheimer's disease (AD) has been studied worldwide, confirming the results (Brogaonkar et al. 1993; Corder et al. 1993, 1994; Mayeux et al. 1993; Noguchi et al. 1993; Payami et al. 1993; Poirier et al. 1993; Roses et al. 1994; Saunders et al. 1993a,b; Strittmater et al.

*Taub Center for Alzheimer's Disease Research in the City of New York. Department of Pathology, Columbia University, New York, NY 10032, USA

A D Roses et al (Eds)
Apolipoprotein E and Alzheimer's Disease

1993a,b; Chartier-Hardlin et al. 1994; Kuusisto et al. 1994; van Duijn et al. 1994; Sorbi et al. 1994; Liddel et al. 1994; Hendrie et al. 1995; Maestre et al. 1995). In a previous study (Maestre et al. 1995) we reported that the association between APOE-ε4 and AD was limited to homozygotes among African-Americans. We also reported that the presence of APOE-ε2 was associated with increased, not decreased, risk as reported by others. The association between APOE-ε4 and AD among Hispanics appeared to be intermediate between that in African-Americans and that in Caucasians. However, studies of other populations including African-Americans have found clear association between APOE-ε4 and AD (Hendrie et al. 1995). With this background, we expanded the case-control study based in the same community to further test the hypothesis that the association between APOE-ε4 and AD differs among African-Americans, Caucasians and Hispanics. We also raise possible explanations for this association.

Methods

Subjects and Setting

We previously described the recruitment of patients and controls into the original investigation (Maestre et al. 1995). We have also previously described the criteria for diagnosis and staging of AD (Hughes et al. 1982; Pittman et al. 1992; Stern et al. 1992). For this investigation, additional cases and controls were identified in a random survey of Medicare recipients in the same community. The Columbia University Institutional Review Board reviewed and approved this project. All individuals provided written informed consent. Both groups of cases and controls were followed over a 4-year period.

Ethnic Group

Ethnic group was classified by self-report using the format of the 1990 United States Census Bureau (Census of Population and Housing 1990). We did not attempt to identify race; rather we allowed each individual to identify himself as belonging to a particular racial or ethnic group. Each subject was also asked whether or not he/she is of "Hispanic" origin, which allowed us to separate subjects into three ethnic groups according to self-report (or informant report): African-American, Hispanic and Caucasian (non-Hispanic).

Genomic DNA Amplification and Restriction Isotyping of APOE

Genomic DNA was amplified by polymerase chain reaction (PCR), using reaction conditions modified from those described by Hixson and Vernier (1991). Briefly, each amplification contained 200 ng genomic DNA, 25 pmol of each primer, 10% dimethyl sulphoxide, and 0.5 U Taq DNA polymerase (Boehringer-Mannheim) in a final volume of 25 μl. Initial denaturation was at of 94°C for 5 min, followed by 40 cycles of annealing at 65°C for 30 s, extension at 72°C for 30 s, denaturation at 94°C for 1 min, followed by a final extension at 72°C for 10 min. The amplification product (10 μl) was then digested with 10 units of Cfol (Boehringer-Mannheim) for 3 h at 37°C and electrophoresed for 15 h at 300 V through a 20% polyacrylamide gel. The genotypes were determined by the sizes of DNA fragments present, viewed and photographed under ultraviolet light after staining with 0.5 τg ethidium bromide/ml. All genotypes were determined without knowledge of patient-control status.

Data Analysis

Allele frequencies for patients with AD and controls were determined by counting alleles and calculating sample proportions. Frequencies of APOE genotypes in patients and controls were compared using the chi square test and the approximate test based on the normal approximation to the binomial distribution (Fleiss 1981). We estimated both simple and stratified (by ethnic group) odds ratios for AD associated with the presence of the APOE-ε4 and ε2 alleles (homozygous and heterozygous), using subjects with the APOE-ε3/ε3 genotype as the reference group. The frequencies for the demographic categories, including ethnic groups, were compared among cases and controls using chi-square analyses and Fisher's exact tests. Tests for homogeneity were also used to measure the degree of variability across the ethnic groups. This statistic has a chi square distribution under the null hypothesis of a consistent (homogeneous) association. Thus, the value of chi is increased when groups are not homogeneous. Both univariate and multivariate odds ratios for AD associated with APOE-ε4 and ε2 were also calculated from logistic regression (Hosmer and Lemeshow 1989), adjusting for age.

Results

Characteristics of Study Population

There were 276 cases and 405 healthy elderly, unrelated controls. Cases and controls differed by gender (AD, 75% women vs. controls, 66% women; X^2, 6.5; $p < 0.01$). In all ethnic groups combined, cases were older (AD 76.4 ± 9.2 years; range, 46–96 years) than controls (72.9 ± 6.7 years; range, 46–93 years; $p < 0.05$). African-Americans and Hispanics were generally older and less well educated than Caucasians (Table 1). Cases also had less education (AD 7.4 ± 5.0 years vs. controls 8.8 ± 4.5 years, $p < 0.05$) than controls.

APOE Allelic Frequencies

To examine the APOE allele frequencies among the three ethnic groups, we used all of the 405 controls and chose a random sample of 10% of cases from each ethnic group to present a "representative" population. The APOE allele frequencies differed significantly in the 3 ethnic groups (X^2, 24.6, 4df; $p <$

Table 1. Demographics and APOE allele frequency among patients with Alzheimer's diesease (AD) and controls, by ethnic groups[a]

Ethnic Groups	Subjects (n)	Mean age (years)	Education (years)	APOE-ε3	APOE-ε4	APOE-ε2
African-American						
AD	95	78 ± 8 4	7 4 ± 4 1	0 63	0 25	0 13
Controls	125	74 ± 5 9[b]	9.1 ± 3.9[b]	0 72	0.21	0 07
Caucasian						
AD	55	71 ± 9.9	13 7 ± 4 1	0 68	0 29	0 03
Controls	74	70 ± 11 6	12 7 ± 3 7[b]	0 87	0 08	0 05
Hispanic						
AD	129	77 ± 8.3	4 9 ± 3 7	0 75	0.19	0 06
Controls	204	73 ± 6 1[b]	7 2 ± 4 3[b]	0.83	0 11	0 06

[a] Allelic frequencies differed significantly between cases and controls within and across ethnic groups See text for explanation
[b] Significant difference within each stratum ($p < 0$ 05)

Table 2. APOE allele frequency by age and ethnic group

Ethnic groups	APOE-ε3	APOE-ε4	APOE-ε2
African-American			
Under age 73	0 72	0 20	0.08
Over age 73[a]	0 67	0 25	0 08
Caucasian			
Under age 73	0.84	0 12	0 04
Over age 73	0.85	0 07	0 08
Hispanic			
Under age 73	0 79	0.13	0 08
Over age 73	0 86	0 09	0.05

To compare ethnic groups across ages, we used all controls recruited and a 10% random sample of cases from each ethnic group to form a representative sample of elderly, age 65 years and older
[a] X^2, 25.1 df4; $p < 0.0001$ APOE allele frequencies differ in African-Americans relative to other ethnic groups over age 73 (median age of the cohort)

0.0001; Table 1). Under age 73 (median age), there were only minimal differences in the APOE allele frequency; African-Americans had slightly higher APOE-ε4 and APOE-ε2 than the other two ethnic groups (Table 2). However, among African-Americans over age 73, the frequency of APOE-ε4 was significantly higher compared to that in Caucasians and Hispanics. APOE allele frequencies did not differ by gender.

Measures of Association: Odds Ratios

There was a strong association between AD and APOE-ε4 homozygosity in all three ethnic groups (Table 3). The odds ratios (OR) for APOE-ε4/ε3 appeared

Table 3. Odds ratios for Alzheimer's disease (AD) associated with APOE genotype by ethnic group

Ethnic groups	ε3/ε3	ε2/ε3	ε2/ε2	ε4/ε2	ε3/ε4	ε4/ε4
African-American						
Alzheimer's disease (n = 95)	40	15	3	2	25	10
Control (n = 126)	60	13	1	3	47	2
Odds ratio	1 0 reference	1 7 (0 8–3 9)	4 5 (0 9–23 4)	1 0 (0 1–7 4)	0 8 (0 4–1 6)	7 5 (2 9–19 7)[b]
Hispanic						
Alzheimer's disease (n = 129)	75	11	1	4	32	6
Control (n = 204)	137	25	1	0	38	3
Odds ratio	1 0 reference	0 8 (0 3–1 9)	1 8 (0 1–29 2)	14 4 (3 5–59 0)[a,b]	1 7 (0 9–4 5)	3 6 (1 1–11 6)[b]
Caucasian						
Alzheimer's disease (n = 55)	26	1	0	2	22	4
Control (n = 74)	55	8	0	1	10	0
Odds ratio	1 0 reference	0 3 (0 02–4 4)	–	4 2 (5 6–31 7)[b]	4 7 (2 2–9 9)[b]	16 9 (3 9–73 0)[a,b]

[a] To calculate odds ratio for "0", we entered '0 5" In this condition we are not able to calculate the true odds ratio Using age-adjusted logistic regression we recalculated the odds ratios, but the results were similar

[b] Statistical significance, $p < 0\ 05$ The odds ratios were significantly different between ethnic groups, as indicated by the test for homogeneity (see text for explanation)

to be stronger for Caucasians (4.4) than African-Americans (0.7) or Hispanics (1.7). Test for heterogeneity of the APOE genotypic association was significant (X^2_H, 25.8, 5df; $p > 0.001$) indicating that the relationships were not parallel.

Although there was a positive association between AD and APOE-ε2/ε3 and APOE-ε2/ε2 in African-Americans and Hispanics, none was statistically significant (Table 3). Among Caucasians the risk of AD associated with these genotypes was reduced.

Because the associations between AD and both APOE-ε4 and APOE-ε2 have been reported to vary with age (Corder et al. 1993; Roses et al. 1994; Roses 1995), we repeated the analysis for all patients with age-at-onset before age 80 and for controls below that age. Without exception the results were very similar, as noted in Table 4. For this analysis we used all genotypes without an APOE-ε4 as the reference to examine the association with APOE-ε4 homozygosity and heterozygosity. As indicated, the association with heterozygosity was less than unity among African-Americans. Again, there was statistical evidence for heterogeneity of the association (X^2_H, 23.2, 2df; $p > 0.001$).

Discussion

As expected, the APOE-ε4/ε4 genotype was most strongly associated with AD in all three ethnic groups. Among Caucasians and, to a lesser extent, Hispanics, APOE-ε4 heterozygosity was also associated with AD, whereas among African-Americans there was an inverse association. There was only minimal evidence of a reduced risk for AD associated with APOE-ε2 genotypes in Caucasians, and there was no decrease in risk for African-Americans or His-

Table 4. Crude odds ratios for Alzheimer's disease (AD) associated with APOE genotypes by ethnic group in individuals with onset before age 80

Ethnic groups	APOE-ε4/ε4	APOE-ε4/ε-	APOE-ε-/ε-
African-American			
Alzheimer's disease	8	11	36
Controls	2	40	66
Odds ratio	7 3 (2 5–21 2)	0 5 (0 2–1 3)	1.0 reference
Hispanic			
Alzheimer's disease	4	24	48
Controls	3	34	143
Odds ratio	3 9 (1 0–16 5)	2 1 (1 1–4.1)	1.0 reference
Caucasian			
Alzheimer's disease	4	21	21
Controls	0	9	50
Odds ratio	19 1 (4 4–83 4)[a]	5 6 (2 5–12 3)	1 0 reference

[a] To calculate odds ratio for "0" we entered "0 5" In this condition we are not able to calculate the true odds ratio Using age-adjusted logistic regression we recalculated the odds ratios, but the results were similar The odds ratios were significantly different between ethnic groups, as indicated by the test for homogeneity (see text for explanation)

panics. In contrast to our earlier report (Maestre et al. 1995), we could not confirm a significant association between APOE-ε2 genotypes and AD. We also examined the odds ratios by including only those individuals whose onset of disease was before age 80, but this did not change the magnitude of the associations in any ethnic group.

Hendrie et al. (1995) previously described a strong association between APOE-ε4 and AD among African-Americans in Indiana. African-Americans were also included in a report by Saunders et al. (1993a), but the specific association between APOE-ε4 and AD was not described. In both studies the APOE-ε4 frequency among controls was lower than that reported here, even in the younger individuals. For example, in Indiana the APOE- ε4 allele frequency was 0.12, whereas in the North Carolina study the allele frequency was 0.17. Differences in APOE allele frequencies among African populations have been described (Sepehrnia et al. 1989; Sandholzer et al. 1995; Zekraoui et al. 1995). A high APOE-ε4 frequency was also observed in South African bushmen (Sandholzer et al. 1995). The APOE-ε4 allele frequency among the African-Americans in this study also did not decline with age, as has been noted in Caucasians (Schachter et al. 1994; Kervinen et al. 1994; Louhija et al. 1994).

The decline in the frequency of APOE-ε4 with advanced age in Caucasians has been attributed to mortality from heart disease (Eichner et al. 1993; Kervinen et al. 1994; Wilson et al. 1994; Luc et al. 1994; Stengard et al. 1995), but this may not affect the association between APOE-ε4 and AD. Finns have an allele frequency of APOE-ε4 that is approximately 0.2 and it decreases with age, yet a strong association between AD and APOE-ε4 remains (Kuusisto et al. 1994). Because the presence of APOE-ε4 alleles does not always influence cholesterol concentrations in African populations (Sepehrnia et al. 1989; Gajra et al. 1994; Sandholzer et al. 1995), it is also possible that the relationship of APOE-ε4 to heart disease, longevity and even AD in African-Americans is not constant.

If APOE-ε4 has a direct effect on AD susceptibility, then African-Americans have either a modifier gene or an environmental factor that "partially" protects them from the effects of the ApoE protein. Clearly, these additional genes or exposures alter the effects of a single APOE-ε4 allele, but not the higher gene dose with homozygosity. Studies to identify genes or exposures that modify the effects of the ApoE proteins will be important to pursue.

Corder et al. (1994) found a negative association between APOE-ε2 and AD in Caucasians. This finding has been a cornerstone for their argument against linkage disequilibrium, because it would require association of APOE-ε4 with an allele increasing susceptibility to AD and association of APOE-ε2 with an allele decreasing susceptibility to AD. We found no such relationship in African-Americans or Hispanics. We originally reported an increased risk associated with APOE-ε2/ε3, but this relationship did not remain statistically significant in the current study, perhaps due to larger number of subjects available. Although we can no longer state that APOE-ε2 is disease associated, we still find no evidence for a "protective" effect among African-Americans or Hispanics.

Because APOE-ε4 heterozygosity (primarily APOE-ε4/ε3) occurs in nearly a quarter or more of the population (Davignon et al. 1988), understanding the protective effect in African-Americans would seem an important point of investigation. It is also possible that genetic variations in other proteins that interact with ApoE directly or indirectly at the cellular level might account for the findings in our study.

Acknowledgments. The authors wish to thank Marcel Kempowski, Adria Gonzales, John Shinin, Maria Gonzales, Harold Brown, Lin Feng, Luwa Yuan and Liz Maestre for their technical assistance in this project. We also acknowledge the efforts of Drs. Barry Gurland, David Wilder and Rafael Lantigua from the Center for Geriatrics and Gerontology at Columbia University in the design and conduct of the Medicare survey. The work was supported by Federal Grants AG07232, AG10963, AG08702 and RR00645 and from the Charles S. Robertson Memorial Gift for Alzheimer's Disease Research from the Banbury Fund.

References

Brogaonkar DS, Schmidt LC, Martin SE, Kanzer MD, Edelsohn L, Growdon J, Farrer LA (1993) Linkage of late-onset Alzheimer's disease with apolipoprotein-E type 4 on chromosome 19 Lancet 342 625

Census of Population and Housing 1990 Summary Tape File 1 (1991) Technical Documentation Prepared by Bureau of Census Washington The Bureau

Chartier-Hardlin M-C, Parfitt M, Legrain S, Pérez-Tur J, Brousseau T, Evans A, Berr C, Vidal O, Roques P, Gourlet V, Fruchart JC, Delacourte A, Rossor M, Amouyel P (1994) Apolipoprotein E, ε4 allele as a major risk factor for sporadic early and late-onset forms of Alzheimer's disease analysis of the 19q13 2 chromosomal region. Human Mol Gen 3 569–574

Corder EH, Saunders AM, Strittmatter WJ, Schmechel DE, Gaskell PC, Small GW, Roses AD, Haines JL, Pericak-Vance MA (1993) Gene dose of apolipoprotein-E type 4 allele and the risk of Alzheimer's disease in late onset families Science 261 921–923

Corder EH, Saunder AM, Risch NJ, Strittmatter WJ, Schmechel DE, Gaskell PC Jr, Rimmler JB, Locke PA, Conneally PM, Schmader KE, Small GW, Roses AD, Haines JL, Pericak-Vance MA (1994) Apolipoprotein E type 2 allele decreases the risk for late onset Alzheimer's disease Nature Gen 7 180–184

Davignon J, Gregg RE, Sing CF (1988) Apolipoprotein E polymorphism and atherosclerosis Arteriosclerosis 8. 1–21

Eichner JE, Kuller LH, Orchard TJ, Grandits GA, McCallum LM, Ferrell RE, Neaton JD (1993) Relation of apolipoprotein E phenotype to myocardial infarction and mortality from coronary artery disease Am J Cardiol 71 160–165

Fleiss JL (1981) Statistical methods for rates and proportions (2nd Edition) John Wiley and Sons, New York, pp 160–187

Gajra B, Candlish JK, Saha N, Mak JW, Tay JS (1994) Effect of apolipoprotein E variants on plasma lipids and apolipoproteins in the Orang Asli ("aborigines") of Malaysia Human Heredity 44 209–213

Hendrie HC, Hall KS, Hui S, Unverzagt FW, Yu CE, Lahiri DK, Sahota A, Farlow M, Musick B, Class CA, Brashear A, Burdine VE, Osuntokun BO, Ogunniyi AO, Gureje O, Baiyewu O, Schellenberg GD (1995) Apolipoprotein-E genotypes and Alzheimer's disease in a community study of elderly African-Americans. Ann Neurol 37 118–120

Hixson J, Vernier D (1991) Restriction isotyping of human apolipoprotein E by gene amplification and cleavage with HhAI J Lipid Res 31 545–548

Hosmer DW, Lemeshow S (1989) Applied logistic regression John Wiley and Sons, New York, pp 38–81

Hughes CP, Berg L, Danizger W, Coben LA, Martin RL (1982) A new clinical scale for the staging of dementia. Brit J Psychiatr 140 566–572

Kervinen K, Savolainen MJ, Salokannel J, Hynninen A, Heikkinen J, Ehnholm C, Koistinen MJ, Kesaniemi YA (1994) Apoliprotein E and B polymorphics – longevity factors assessed in nonagenarians Atherosclerosis 105· 89–95

Kuusisto J, Koivisto K, Kervinen K, Mykkanen L, Helkala EL, Vanhanen M, Hanninen T, Pyorala K, Kesaniemi YA, Riekkinen P, Laakso M (1994) Association of apolipoprotein-E phenotypes with late onset Alzheimer's disease population based study Brit Med J 309: 636–638

Liddell M, Williams J, Bayer A, Kaiser F, Owen M (1994) Confirmation of association between the ε4 allele of apolipoprotein E and Alzheimer's disease J Med Gen 31 197–200

Louhija J, Miettinen HE, Kontula K, Tikkanen MJ, Miettinen TA, Tilvis RS (1994) Aging and genetic variation of plasma apolipoprotiens Relative loss of the apolipoprotein E4 phenotype in centenarians Arteriosclerosis Thrombosis 14. 1084–1089

Luc G, Bard JM, Arveiler D, Evans A, Cambon JP, Bingham A, Amouyel P, Schaffer P, Ruidavets JB, Cambien F, Fruchart JC, Ducimetière P (1994) Impact of apolipoprotein E polymorphism on lipoproteins and risk of myocardial infarction The ECTIM Study Arteriosclerosis Thrombosis 14. 1412–1419

Maestre G, Ottman R, Stern Y, Gurland B, Chun M, Tang MX, Shelanski M, Tycko B, Mayeux R (1995) Apolipoprotein-E and Alzheimer's disease ethnic variation in genotypic risks. Ann Neurol 37 254–259

Mayeux R, Stern Y, Ottman R, Tatemichi TK, Tang MX, Maestre G, Ngai C, Tycko B, Ginsberg H (1993) The apolipoprotein ε4 allele in patients with Alzheimer's disease Ann Neurol 34 752–754

Noguchi S, Murukami K, Yamada N (1993) Apolipoprotein E genotype and Alzheimer's disease Lancet 342 737

Payami H, Kaye J, Heston LL, Bird TD, Schellenberg GD (1993) Apolipoprotein-E genotype and Alzheimer's disease Lancet 342 738

Pittman J, Andrews H, Tatemichi T, Link B, Struening E, Stern Y, Mayeux R (1992) Diagnosis of dementia in a heterogenous population a comparison of paradigm-based diagnosis and physician's diagnosis Arch Neurol 49. 461–467

Poirier J, Davignon J, Bouthillier D, Kogan S, Bertrand P, Gauthier S (1993) Apolipoprotein E polymorphism and Alzheimer's disease Lancet 342 697–699

Roses AD (1995) Apolipoprotein E genotyping in the differential diagnosis, not predoctop of Alzheimer's disease Ann Neurol 38 6–14

Roses AD, Strittmatter WJ, Pericak-Vance MA, Corder EH, Saunders AM, Schmechel DE (1994) Clinical application of apolipoprotein E genotyping to Alzheimer's disease Lancet 343· 1564–1565

Sandholzer C, Delport R, Vermaak H, Utermann G (1995) High frequency of the apo ε4 allele in Khoi San from South Africa Human Gen 95 46–48

Saunders AM, Schmader K, Breitner JCS, Benson MD, Brown WT, Goldfarb L, Goldgaber D, Manwaring MG, Szymanski MH, McCown N, Dole KC, Schmechel DE, Strittmatter WJ, Pericak-Vance MA, Roses AD (1993a) Apolipoprotein-E ε4 allele distributions in late-onset Alzheimer's disease and in other amyloid-forming diseases Lancet 342 710–711

Saunders AM, Strittmatter WJ, Schmechel D, St George-Hyslop PH, Pericak-Vance MA, Joo SH, Rosi BL, Gusella JF, Crapper-Maclachan DR, Alberts MJ, Hulette C, Crain B, Goldgaber D, Roses AD (1993b) Association of apolipoprotein E allele ε4 with late-onset familial and sporadic Alzheimer's disease Neurology 43 1467–1472

Schachter F, Faure-Dalanef L, Guenot F, Rouger H, Froguel P, Lesueur-Ginot L, Cohen D (1994) Genetic associations with human longevity at the APOE and ACE loci Nature Gen 6 29–32

Sepehrnia B, Kamboh MI, Adams-Campbell LL, Bunker CH, Nwankwo M, Majumder PP, Ferrell RE (1989) Genetic studies of human apolipoproteins X The effect of the apolipoprotein E polymorphism on quantitative levels of lipoproteins in Nigerian blacks Am J Hum Genet 45 586–591

Sorbi S, Nacmias B, Forleo P, Latorraca S, Gobbini I, Bracco L, Piacentini S, Amaducci L (1994) ApoE allele frequencies in Italian sporadic and familial Alzheimer's disease Neurosci Lett 177 100–102

Stengard JH, Zerba KE, Pekkanen J, Ehnholm C, Nissinen A, Sing CF (1995) Apolipoprotein E polymorphis predicts death from coronary heart disease in a longitudinal study of elderly Finnish men Circulation 91 265–269

Stern Y, Andrews H, Pittman J, Sano M, Tatemichi T, Lantigua R, Mayeux R (1992) Diagnosis of dementia in a heterogeneous population development of a neuropsychological paradigm and quantified correction for education Arch Neurol 49 453–460

Strittmatter WJ, Saunders AM, Schmechel D, Pericak-Vance MA, Enghild J, Salvesen GS, Roses AD (1993a) Apolipoprotein E high affinity binding to beta-amyloid and increased frequency of type 4 allele in late-onset familial Alzheimer's disease Proc Natl Acad Sci USA 90 1977–1981

Strittmatter WJ, Weisgraber KH, Huang D, Dong L-M, Salvesen GS, Pericak-Vance M, Schmechel D, Saunders AM, Goldgaber D, Roses AD (1993b) Binding of human apolipoprotein E to β-A4 peptide isoform-specific effects and implications for late-onset Alzheimer's disease Proc Natl Acad Sci USA 90 8098–8102

Van Duijn CM, de Knijff P, Cruts M, Wehnert A, Havekes LM, Hofman A, Van Broeckhoven C (1994) Apolipoprotein E4 allele in a population based study of early-onset Alzheimer's disease Nature Gen 7 74–78

Wilson PW, Myers RH, Larson MG, Ordovas JM, Wolf PA, Schaefer EJ (1994) Apolipoprotein E alleles, dyslipidemia, and coronary heart disease The Framingham Offspring Study JAMA 272. 1666–1671

Zekraoui L, Lucotte G, Raisonnier A (1995) Apolipoprotein E genotypes in Africa· alleleic frequencies and ethnic groups Paper presented at the IPSEN Foundation Meeting Paris, France, May 29

The Search for Additional Alzheimer's Disease Genes

M.A. Pericak-Vance, P.M. Conneally, G.W. Small, A.M. Saunders, L. Yamaoka, P.C. Gaskell, C. Robinson, M. Ter-minassian, P.A. Locke, M. Pritchard, C.S. Haynes, J. Growdon, J.F. Gusella, A.D. Roses,* and *J.L. Haines*

Summary

Alzheimer disease (AD) is a complex, heterogeneous, genetic disorder that has resisted most attempts to understand its genetic etiology. Fortunately, linkage analysis has proven to be both powerful and successful in unraveling the genetic components of AD. Three loci have been identified; the amyloid precursor protein (APP) on chromosome 21, a newly identified gene on chromosome 14 and the apolipoprotein E locus (APOE) on chromosome 19. These loci account for approximately one half of the genetic etiology of AD. We have undertaken a genomic screening effort to identify additional AD genetic effects. Using a multi-analytical approach we have examined over 175 marker loci for linkage with AD in a series of late-onset (onset after 60 years of age) families. The results of these analyses have shown three chromosomal regions that could potentially contain additional AD susceptibility loci. These regions of interest will be further evaluated using additional markers and AD families to confirm these initial findings.

Introduction

Alzheimer's disease (AD) is a complex genetic disorder. Although it had long been recognized that there are relatively rare early-onset (onset before 60 years of age) families that segregate a Mendelian form of AD, less than a decade ago it was still questioned whether or not genetics played any role in the late-onset (onset after 60 years of age) form of the disease. In the past eight years significant progress has been made in unraveling the complex etiology of this heterogeneous disorder using the powerful tools of genetic linkage analysis. To date, three loci have been identified as contributing to the genetic etiology of AD. These include the amyloid precursor protein (APP) on chromosome 21 (Goate et al. 1991), a newly identified gene on chromosome 14 (Presenilin I) (Sherrington et al. 1995) and the apolipoprotein (APOE) locus

* Departments of Medicine (Neurology) and Genetics, Joseph and Kathleen Bryan Alzheimer's Disease Research Center, P O Box 2900, Duke University Medical Center, Durham, NC 27710, USA

A D Roses et al (Eds)
Apolipoprotein E and Alzheimer's Disease

on chromosome 19 (Pericak-Vance et al. 1991; Strittmatter et al. 1993). The APP and chromosome 14 locus contribute to early-onset AD, whereas APOE is generally associated with late-onset AD. The effect of APOE on susceptibility in AD has been confirmed in multiple racial groups and ethnic populations worldwide. The APOE-4 allele acts to increase risk and decrease the age-of-onset distribution in AD (Corder et al. 1993). These three loci (APP, Presenilin I, APOE) account for about half of the total genetic etiology in AD (Risch et al. 1994); thus, there are still AD loci that need to be identified. This is particularly well highlighted by late-onset families with pathologically confirmed AD who do not segregate the APOE4 allele.

To further dissect the genetics of late-onset AD we have undertaken a genomic screen using a tiered multi-analytical approach that combined both model-dependent and model-independent methods of linkage analysis. The finding of these additional AD loci will foster new studies into the etiology of AD, including the examination of interaction among loci as well as the interaction between genetics and potential environmental causes. These studies will eventually allow us to unravel the mystery of this complex genetic disease.

Materials and Methods

Family Data

We have ascertained and sampled over 155 late-onset AD (onset after 60 years of age) families with a total of 415 sampled AD affected individuals. All sampled individuals diagnosed with probable AD were examined by a neurologist or associated personnel at the Joseph and Kathleen Bryan Alzheimer's Disease Research Center (ADRC) at Duke University, the Massachusetts General Hospital ADRC, the University of California, or the Indiana Alzheimer Disease Center's National Cell Repository. The clinical diagnosis was made using the NINCDS-ADRDA diagnostic criteria (McKhann et al. 1984).

A subset of 16 families with 142 sampled individuals (44 sampled affecteds) was selected to rapidly complete a total genomic screen. These families were chosen because they had a large number of sampled affected individuals in each pedigree and did not have any APOE-4/4 affected individuals. The included families were also negative for the APP mutation and showed no evidence of linkage to the chromosome 14 locus (St George-Hyslop et al. 1992). All families in the screen also had one or more autopsy-confirmed AD cases. An example pedigree is shown in Fig. 1.

DNA Analysis

Genomic DNA was extracted from whole blood using methods described elsewhere (Pericak-Vance et al. 1991; Locke et al. 1995). APOE and marker genotypings were performed according to standard techniques as previously

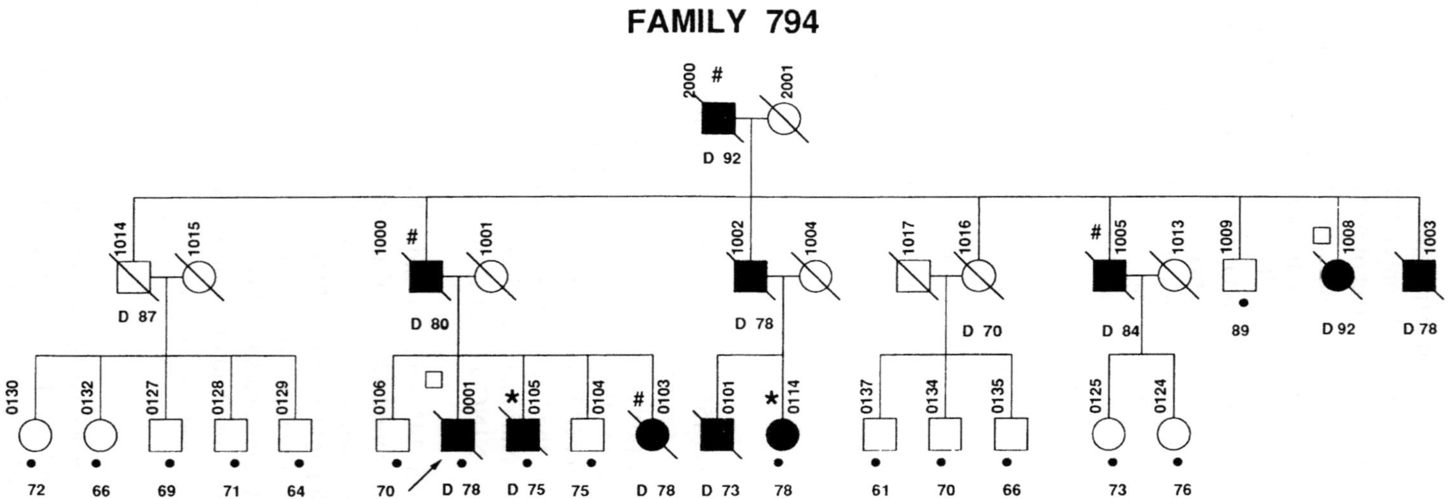

Fig. 1. Example of a late-onset Alzheimer's disease (AD) pedigree used in the linkage analysis *Black symbols*, affected by AD Numbers after *D* beneath *crossed symbols* indicate age at death, numbers beneath *uncrossed symbols* indicate age at examination, *black dots* indicate DNA, *open squares* above symbols indicate a confirmed autopsy, * indicate a clinical diagnosis of AD

described (Saunders et al. 1993; Yamaoka et al. 1994). Paternity was confirmed by examining a series of linked polymorphic markers (haplotypes) over several chromosomal regions. Microsatellite marker loci were chosen for analysis based on heterozygosity (HET > 0.70) using a 10–20cM template for the initial genomic screen.

Statistical Analysis

The families were analyzed for linkage using a multi-analytical approach that employed the use of both model-dependent (lod score) (OH, 1991) and model independent (affected pedigree member analysis, APM) linkage methods (Weeks and Lange 1988). The LINKAGE Program Package (version 5.1) was used in the lod score analysis (Lathrop and Lalouel 1984). For the lod score approach, autosomal dominant inheritance was assumed and both age-dependent and low penetrance (affecteds only) analyses were performed on each marker locus. The age curve used in the analysis for assigning probabilities of affection status to at-risk families members was generated assuming a normal distribution of age of onset with a sample mean and variance of 66.1 ± 10.3 years. For the low penetrance analysis, although genotypic data were included on all sampled individuals, AD phenotypic data were limited to affected individuals. The gene frequency assumed for the AD locus was 0.001 for the AD allele. Marker allele frequencies were estimated using a series of 50–100 unrelated Caucasian controls together with the unrelated married in AD family members. The APM statistic was applied as described previously, with the test statistic of $p = 1/\mathrm{sqrt}[p]$ used in evaluating the APM significance.

The critical values considered for follow-up analysis were a nominal p value of 0.05 or less for APM analysis and a lod score of 1.00 or more for either the age-dependent or the low penetrance lod score analysis. Using these criteria a region was marked as "interesting" if it exceeded the critical values on both the APM and lod score analysis for a single marker or exceeded the critical value on either APM or lod score analyses for two (or more) markers in the same region (Table 1). Markers were considered to be in the same region if the genetic distance between them was given as less than 30cM.

Results

We have genotyped over 175 marker loci covering approximately 2800cM, which represents approximately 75% of the total genome. The findings for these markers are summarized in Table 2. Of the markers genotyped and analyzed, only four chromosomal regions reached our criteria for initial follow-up analysis using additional markers and families. Region 1 has three markers with a lod score over 1.00 for both the age-adjusted (AJ) and the low

Table 1. Alzheimer's disease genomic screen criteria for follow-up analysis

Critical values	
APM· nominal p value $\leqslant$0 05	
LOD score $Z > 1\ 00$, any model	
Criteria for follow-up	
Exceed critical values on both APM and lod score analyses for at least one marker	
Exceed critical values on either APM or lod score analysis for two (or more) adjacent markers (approximately 30 cM region)	

APM, affected pedigree member analysis, lod, logarithm of the odds of linkage

Table 2. Frequency of positive results

	Single marker	Regions[a]
APM analysis ($p \leqslant 0\ 05$)	20	16
Lod score analysis (any model Z $\geqslant$ 1 00)	5	4

Four regions qualified for follow-up analysis
[a] Genetic map distance of 30 cM or less

penetrance (LP) analysis ($\hat{z}_{LP}$, 1.75, at θ, 0.15), as well as significant APM results for several adjacent makers ($p < 0.001$). Regions 2 and 3 have significant APM results ($p < 0.001$) on several adjacent markers. These regions also have positive lod scores ($\hat{z}_{LP}$, 0.60 on $\hat{z}_{LP}$, 1.0, respectively). Region 4 has significant APM ($p < 0.001$) and a positive lod score greater than 1.00 on both the low-penetrance and age-adjusted anlaysis ($\hat{z}_{LP}$, 2.36; $\hat{z}_{AJ}$, 1.76; θ, 0.0).

Discussion

The ability to identify additional AD genetic factors and then examine their interaction with both APOE and potential environmental factors is critical to fully dissecting the etiology of this complex disease. Our studies (Roses et al. 1995) have shown that APOE accounts for at most 50% of the total genetic effect in AD. This leaves a substantial amount of genetic variation still unexplained. Linkage analysis is a powerful approach for unraveling this remaining genetic effect.

As AD is a complex disorder, we have chosen a multi-analytical approach to this problem that employs linkage techiques of both a model-dependent and model independent nature. Our first pass through the genome is not intended to identify with certainty "the" new AD regions but is structured to highlight regions of interest for follow-up analysis. Thus our critical values chosen for the follow-up criteria (LOD $\geqslant$ 1.00, nominal APM p value = 0.05) are not being used to establish linkage but to ensure that our pobability of type 2 error (saying that there is no linkage when there is in fact linkage) is relatively low.

We have also tiered our screening families so that our initial pass through the genome is performed on AD families that do not appear to segregate the APOE-4 effect. In retrospect, these families appear to be some of the larger families available to us for study in terms of power to detect linkage.

Our initial analysis involving 75% of the genome identified two regions warranting an initial follow-up analysis. We must stress that we do not feel that the present results are proof that a susceptibility gene definitively lies in one of these regions but only suggest that possibility. Follow-up analyses will include multipoint linkage analysis of markers in those regions, genotyping of additional AD families and genotyping of additional markers in the regions. Once a particular linkage is confirmed we will examine the region for possible candidate gene loci as well as examine the interaction of APOE with the new candidate gene area, employing such techniques as two locus linkage analysis.

In summary, we have completed a genomic screen spanning 75% of the genome in an effort to identify additional AD susceptibility loci. The data have been analyzed using a multi-analytical approach that employs several methods of analysis. Based on our apriori criteria, we have identified two regions of the genome that are targeted for our initial follow-up analysis. These analyses are now ongoing and include the testing of additional markers and families in the regions of interest.

Acknowledgments. This work is supported by grants from the NINCDS 5RO1NS31153 (MP-V, JLH), NIA LEAD Award 5R35 AGO7922 (ADR), MH 46424(GNS) NIA P30AG10133 (PMC) AG11268 (MP-V) and a Zenith award grant from the Alzheimer's Disease and Related Disorders Association (JLH), 11RG-94-101 from the Alzheimer and Related Disorders Association (GWS). We would also like to thank the many family members and patients with AD for participating in this study.

References

Corder EH, Saunders AM, Strittmatter WJ, Schmechel DE, Gaskell PC, Small GW, Roses AD, Haines JL, Pericak-Vance MA (1993) Gene dose of apolipoprotein E type 4 allele and the risk of Alzheimer's disease in late onset families Science 261 921–923

Goate A, Chartier-Harlin M-C, Mullan M, Brown J, Crawford F, Fidani L, Giuffra L, Haynes A, Irving N, James L, Mant R, Newton P, Rocke K, Roques P, Talbot C, Pericak-Vance M, Roses A, Williamson R, Rossor M, Owen M, Hardy J (1991) Segregation of a missence mutation in the amyloid precursor gene with familial Alzheimer's disease Nature 349: 704–706

Lathrop GM, Lalouel JM (1984) Easy calculations of lod scores and genetic risks on small computers Am J Human Genet 36: 460–465

Locke P, Conneally PM, Tanzi RE, Gusella JF, Haines JL (1995) APOE and Alzheimer disease examination of allelic association and effect on age at-onset in both early and late onset cases Genet Epidemiol 12. 83–92

McKhann G, Drachman D, Folstein M, Katzman R, Price D, Stadlan EM (1984) Clinical diagnosis of Alzheimer's disease· report of the NINCDS-ADRDA work group under the auspices of department of health and human services task force on Alzheimer's disease. Neurology 34: 939–944

Oh J (1997) Analysis of human genetic linkage. The Johns Hopkins University Press, Baltimore London

Pericak-Vance MA, Bebout JL, Gaskell PC, Yamaoka LH, Hung W-Y, Alberts MJ, Walker AP, Bartlett RJ, Haynes CA, Welst KA, Earl NL, Heymark A, Clark CM, Roses AD (1991) Linkage studies in familial Alzheimer's disease evidence for chromosome 19 linkage Am J Hum Genet 48 1034–1050

Risch NJ et al. (1994) Presentation The 1994 Adler Conference on Alzheimer Disease. Duke University, Durham, NC

Roses AD, Devlin B, Conneally PM, Small GW, Saunders AM, Pritchard M, Locke PA, Haines JL, Pericak-Vance MA, Risch N (1995) Measuring the genetic contribution of APOE in late-onset alzheimer disease (AD) 45th Annual American Society of Human Genetics Meeting, Minneapolis, MN

Saunders AM, Strittmatter WJ, Schmechel D, St George-Hyslop PH, Pericak-Vance MA, Joo SH, Rosi BA, Guesella JF, Crapper-MacLachlan DR, Alberts MJ, Hulette C, Crain B, Goldgaber D, Roses AD (1993) Association of apolipoprotein E allele ε4 with late-onset familial and sporadic Alzheimer's disease Neurology 43 1467–1472

Sherrington R, Rogaev EI, Liang Y, Rogaeva EA, Levesque G, Ikeda M, Chi H, Lin C, Li G, Holman K, Tsuda T, Mar L, Foncin J-F, Bruni AC, Montesi MP, Sorbi S, Rainero I, Pinessi L, Nee L, Chumakov I, Pollen D, Brookes A, Sanseau P, Polinsky RJ, Wasco W, Da Silva HAR, Haines JL, Pericak-Vance MA, Tanzi RE, Roses AD, Fraser PE, Rommens JM, St George-Hyslop PH (1995) Cloning of a gene bearing missense mutations in early-onset familial Alzheimer's disease Nature 375· 754–760

St George-Hyslop PH, Haines J, Rogaev E, Mortilla M, Vaula G, Pericak-Vance M, Foncin J-F, Montesi M, Bruni A, Sorbi S, Rainero I, Pinessi L, Pollen D, Polinsky R, Nee L, Kennedy J, Macciardi F, Rogaeva E, Liang Y, Alexandrova N, Lukiw W, Schlumpf K, Tanzi R, Tsuda T, Farrer L, Cantu J-M, Duara R, Amaducci L, Bergamini L, Gusella J, Roses A, Crapper-McLachlan D (1992) Genetic evidence for a novel familial Alzheimer's disease locus on chromosome 14, Nature Genet 2 330–334

Strittmatter WJ, Saunders AM, Scmechel D, Pericak-Vance MA, Enghild J, Salvesen GS, Roses AD (1993) Apolipoprotein E. high avidity binding to β-amyloid and increased frequency of type 4 allele in late-onset familial Alzheimer disease Proc Natl Acad Sci USA 90 1977–1981

Weeks DE, Lange K (1988) The affected-pedigree-member method of linkage analysis Am J Human Genet 42 315–326

Yamaoka LH, Westbrook CA, Speer MC, Gilchrist JM, Jabs EW, Schweins EG, Stajich JM, Gaskell PC, Roses AD, Pericak-Vance MA (1994) Development of a microstatellite genetic map spanning 5q31–q33 and subsequent placement of the LGMD1 locus between D5S178 and IL9 Neuromuscular Diseases 4 471–475

Subject Index

acetylcholine 22, 26
actin 113
ADAS score 24
additional Alzheimer's disease genes 180–186
advanced glycosylation end products (AGEs) 168
aging 29–48, 59–73, 104–106, 129, 133, 161
amyloid (Aβ) 2–7, 11, 12, 15, 16, 21, 22, 29, 32, 34, 35, 40–42, 44, 45, 50, 56, 59, 60, 69, 74–77, 91, 92, 97–100, 103–107, 130, 136–144, 147, 151–160, 162
amyloid precursor protein (APP) 2, 3, 5, 75, 92, 180, 181
amyloidogenesis 31, 34, 40, 41, 42, 44, 75, 76, 92, 98, 152, 153, 158
antisense 113
apoE/Aβ interactions 3, 4, 6, 7, 12, 15–17, 21, 22, 50, 56, 76, 91, 97, 99, 136–142, 151–158
apoE and cognitive processes 22, 25, 26
apoE immunoreactivity during brain aging 29–48
apoE knockout mice 22, 25, 29, 32–34, 59–70
apoE receptor 5, 20, 21, 45, 70, 74–77, 92, 97–102
apolipoprotein A 64, 156
apolipoprotein B 20, 21, 89
apolipoprotein J 151, 152
astrocyte 30, 37, 44, 45, 60, 66, 68, 70, 74, 76, 77, 81–84, 91, 98, 100, 137, 141
atherosclerosis 137

baculovirus 113
bimolecular interaction analysis 147–149

calcium 60, 69, 112, 117
calcium-binding protein 60
calmodulin 112, 117
cellular uptake of apoE-containing lipoproteins 136–144
CERAD criteria 41
chaperone 98, 151–160
cholesterol 11, 14, 20, 22, 29, 30, 44, 50, 51, 60, 69, 75, 78, 91, 97, 136–138
choline 22
choline acetyl transferase 22–24, 32, 36, 37
cholinergic injury 29, 36–41
cholinergic integrity 20–28
cholinergic system 20–28, 37–38
cholinesterase inhibitor (cf also tacrine) 24, 26
chromosome 14 2, 3, 5, 51, 104, 180, 181
chromosome 19 2, 5, 12, 19, 21, 30, 50, 51, 104, 180, 181
chromosome 21 2, 51, 104, 180
clinical-pathological correlates 98, 99
cognition 7, 24–26, 98, 103, 104
congophilic angiopathy 99
cyclin-dependent kinase 5 (cdk5) 111, 112, 117, 129
cytoskeleton 6, 16, 49–60, 69, 70, 119

dendrite 6, 7, 22, 31, 36, 41, 42, 59, 61–63, 65, 67
dendrite remodelling 20
development (and tau protein) 109–114, 131
diagnosis 5
Down's syndrome 31

early-onset Alzheimer's disease 2, 3, 5, 51, 104, 181
electron microscopy 6, 7
endocytosis 75, 91, 92, 137, 142
entorhinal cortex lesion 21, 63, 66, 69
ethnic variation 5, 170–179, 181

free radicals 60, 161–168
function of apoE 49–73

genetic 5–7, 11, 12, 17, 97, 98, 100, 103, 104, 170–186
glial cell 29–31, 36, 37, 40, 44, 45, 65, 66, 68, 70, 74, 81, 83, 137

glycogen synthase kinase-3 (GSK3) 111, 112, 117, 119, 129

heparan sulfate proteoglycan 91
heterogeneity of Alzheimer's disease 2, 12, 104, 180, 181
HDL 14, 15, 21, 50
hippocampus 20–23, 29, 30, 34, 37, 39, 41, 44, 59, 61, 63, 75, 79, 80, 82, 85–88, 99, 103, 107
history 1, 2
hyperlipoproteinemia 13, 50

integrity of the aging central nervous system 59–73
interactions of apoE with microtubule-associated proteins 145–150
intracellular trafficking 4, 6

late-onset Alzheimer's disease 2, 3, 5–7, 21, 51, 145, 181, 182
LDL 137
LDL receptor 5, 13, 20–22, 50, 52, 55, 60, 74, 76–78, 89, 91, 97, 99, 136, 137, 139–141, 146, 147
LDL-receptor related-protein (LRP) 5, 49, 50, 52, 54, 55, 74–77, 89–91, 97–100, 136–144, 146
linkage 2, 5, 7, 12, 176, 181–185
localization of apoE in the neuron 3–7, 44, 45, 56
Lod score 7, 183
loss of synapses 22, 60–62, 69, 75, 104

MAP2 6, 7, 45, 61–65, 70, 81, 83, 113, 146
MAP kinase 111, 112, 114, 117, 129
MAR kinase 126, 129, 131, 132
membrane alteration in Alzheimer's disease 75
memory 7, 8, 25, 103
microglial cell 36, 37, 45, 100
microtubule 4, 7, 16, 49–56, 70, 76, 104, 109, 111–116, 126–132, 141, 145–150
model of apoE structure 13
myelin 75, 98, 137

neurofibrillary tangles 2, 3, 5, 11, 12, 16, 21, 30, 44, 50, 51, 69, 70, 75, 76, 86, 98, 99, 103–135, 137, 146, 149
neuronal death 7, 36, 75, 104
neuronal injury and apoE 11, 20, 29, 30, 44, 50, 59, 60, 63–70, 75, 76, 100, 137, 146
neuronal outgrowth (effect of apoE on) 6, 11, 17, 49–58, 98, 113, 114
neuropathological stages of Alzheimer's disease 106, 107
neurotrophic factor 60, 63
nicotinic receptor 23, 24

oxidation 129–133, 161–169

paired helical filament (cf. also neurofibrillary tangles) 5, 16, 51, 76, 104–108, 114–117, 120, 126–135, 146, 147, 149
pathogenesis 5, 49, 89–92, 103, 104, 119, 120, 136, 145, 149, 151
phosphatase 112, 117, 119
phospholipids 22
phosphorylation of tau 4, 16, 50, 51, 70, 76, 104, 109–120, 126–132, 146, 147, 149
prediction 5
prosimian 29–49
PrP 158

S182 gene 3, 5
senile plaque 2, 3, 11, 12, 21, 29, 30, 34–36, 40, 56, 75, 76, 86, 88, 90, 91, 100, 103, 137, 146, 151, 152, 156, 158, 162
structural analysis 6, 11–19
surface plasmon resonance technique 6, 147, 148
synaptic plasticity and remodelling 11, 20–30, 44, 45, 51, 59–73, 76
synaptophysin 6, 59, 61, 63–65

tacrine 7, 24–26
tau 4–7, 12, 16, 17, 50, 51, 70, 76, 103–135, 141, 146–150
tau function 112–114
tau-microtubule interaction 127–129
therapy 4, 5, 7, 8, 17, 24–26, 100, 120, 132
three-dimensional structures of apoE 14, 15
transgenic mice 5, 29, 31, 118–120
tubulin 4, 6, 49, 54, 59, 63, 78, 113, 146, 147

ubiquitin 115, 116, 129

VLDL 14, 15, 21, 50–55, 77, 136–142
VLDL receptor 5, 6, 74–97, 99, 100, 146

X-ray cristallography 12–15

Printing Saladruck, Berlin
Binding Buchbinderei Lüderitz & Bauer, Berlin

MIX
Papier aus verantwortungsvollen Quellen
Paper from responsible sources
FSC® C105338

If you have any concerns about our products,
you can contact us on
ProductSafety@springernature.com

In case Publisher is established outside the EU,
the EU authorized representative is:
Springer Nature Customer Service Center GmbH
Europaplatz 3, 69115 Heidelberg, Germany

Printed by Libri Plureos GmbH
in Hamburg, Germany